Move On with PARKINSON'S

Michael Stanfield

AN INSPIRING TRUE STORY
AS TOLD BY A PD PATIENT

NOTICE: The information in this book is for educational purposes only and it does not constitute medical advice nor should it be construed as such. The effectiveness and safety of any drug, treatment or advice mentioned in this book cannot be guaranteed. Some of the tips may not be effective for everyone. A good physician is the best judge of proper medical treatment needed for certain conditions. We recommend that you consult with your physician or health care professional before taking or discontinuing any medication or changing your treatment in any way.

Designed and manufactured
in the United States of America

Published by Lulu.com
Lulu ID: 2405978
www.lulu.com
for information contact Lulu.com
3131 RDU Center Dr., Suite 210
Morrisville NC 27560

ISBN 978-0-557-06499-1

You can contact the author at jcstanfld@aol.com

ACKNOWLEDGEMENTS

I wish to thank my wife Jeanne for the countless hours of work and support that she has devoted to helping me deal with Parkinson's disease.

I appreciate the assistance and concern offered by my family, son Mike and his wife Stacey, son Christopher and his girlfriend Monica, and daughter Betsy. I also thank my sister Clara for her help and for being one of the first to detect the onset of my illness.

It is with profound gratitude that I acknowledge the medical skills and knowledge of Dr. Jay Rao, MD, Ochsner Clinic in New Orleans. His uncanny ability to diagnose and treat Parkinson's disease has made my life infinitely better.

I am grateful for the contribution of Ms. Maribel Bleeker, a professional Health Fitness Specialist, who designed and implemented an exercise program that has restored much of my strength, balance, and coordination.

For the Patients of Parkinson's Disease—

as we undertake this long journey,
remember that . . .

you are not alone

Contents

PART THREE

TIPS FOR LIVING WITH PD

APPENDICES

PROLOGUE

When I was first diagnosed with Parkinson's disease, I became very upset and discouraged. Although I knew virtually nothing about PD, I told Jeanne that I considered my life to be essentially over. I am one really lucky guy—my wife is a former registered nurse, she has been around people who are gravely ill, she is a "take-charge" kind of person, and she is loyal—someone who will be there for you in a pinch.

One of the first things she did was to visit the local library in order for us both to learn something about the disease. She came home with an armful of books, some by medical doctors, other books by patients. Most of the books upset me because the authors described terrible symptoms and offered little encouragement. But books don't treat diseases, doctors do.

Jeanne had already begun making contacts to find a specialist for Parkinson's disease. Fortunately she located a movement disorder specialist who could see us right away. He is Dr. Jay Rao in New Orleans, Louisiana. My initial visit accomplished two important things: first, I was promptly put under medical treatment; and second, I learned that I was not alone—superb help was on the way. I regained my composure and vowed to fight the disease and win. It no longer mattered what the books said. I was ready to do battle on my own terms. A two-part treatment plan was begun that included medication and an intensive exercise program.

That was almost five years ago. I am pleased to report that so far I have held my own against PD, and I feel that I have actually turned the clock back so that my symptoms are better than at the start of treatment. As the title suggests, I am moving on in spite of Parkinson's disease. My life has begun again, I am good again.

Dr. Rao encouraged me to tell my story, which is mostly positive, to inspire and educate other PD patients. I have decided to do just that, hence the book at hand. After reviewing existing books and literature on Parkinson's disease, I concluded that there are many excellent books covering a host of subjects related to the disease, such as technical reviews of neurological and medical topics, papers about scientific studies of PD, patient accounts of their own experiences, and descriptions of new treatments and drugs. The message that is central to my story is "*YOU ARE NOT ALONE.*"

The book is in three parts. ***Part One*** tells how my life has changed with PD. ***Part Two*** tells my story chronologically. And ***Part Three*** offers some tips to help the reader address his or her situation from the standpoint of hope, not despair.

Be courageous. You are on a new "life path." You can move on with Parkinson's. And don't forget the power of laughter. Be of good spirits. Together we will win.

HUMOR IS NO JOKE

Parkinson's disease is no laughing matter. Well, wait a minute. Yes, sometimes it is. Maribel Bleeker, my personal trainer, has me perform facial exercises every session, which include a GREAT BIG smile while looking into a mirror. To me, I look comical and I cannot resist laughing. As my facial muscles become stronger I can once again smile at people whom I meet. A dermatologist recently complimented me on how good my facial expressions look. He said that his brother had PD and he had a vacant, expressionless face. I can also frown and look surprised. Sometimes I stick out my tongue, not to be rude, but to keep my tongue working. My trainer can't help laughing with me.

We joke a lot. Almost anything has a funny side to it. Each exercise session is planned in advance and documented in a written record. Maribel brings to the thrice-weekly one-hour session whatever exercise equipment is needed, such as weights, balance pads, etc. As she arrived recently I joked that I had a vision that she was bringing an orange traffic cone, and sure enough she had one.

She replied, "Oh, it's for you to wear on your head, after all, you *are* a 'Conehead' aren't you?" She was referring to a sketch on *Saturday Night Live* about a hilariously funny family from another planet. Ha-ha-ha. And another session starts on a light note.

NOTE FROM THE AUTHOR

Telling the story of my experiences with Parkinsons's disease is complicated by the passage of time and the ever-changing variety of treatments available. Where should I begin? How much detail should I include? I have scientific and engineering training and two degrees from MIT. Should the book be long and technical? Or should it be short and colloquial, like a human interest cover story in a Sunday newspaper supplement, where entertainment is the goal?

I chose a middle ground approach, beginning in Part One with a broad summary of how Parkinson's disease has affected my life.

Part One

How Parkinson's Disease Changed my Life

It does not matter how slowly you go
so long as you don't stop.

Confucius

—1— DIAGNOSIS

LEAD PIPES DON'T BEND

When a doctor moves your arm around and lifts it by the wrist or elbow he or she is checking on its rigidity or stiffness. If your arm does not bend freely, especially at the elbow, you may have one of the key symptoms of Parkinson's disease.

Jeanne says she noticed that I walked stiffly without swinging my arms, and my right elbow was bent slightly. This description is practically a text book sign of Parkinson's disease. When she called my attention to the stiff, bent elbow, I looked down, straightened my right arm and elbow, and dismissed her observation as unimportant.

While I did not see any significance at the time, I was surprised to see myself in a mirror after the diagnosis of PD. What was most unusual was that I was completely unaware of my rigidity before it was brought to my attention.

My arms and legs felt useless, heavy, non-responsive. The more I fought against the stiffness, the worse it seemed to be.

Drug therapy and exercise can help reduce rigidity and avoid loss of use of the limbs.

I never think of the future - it comes soon enough.

Albert Einstein

—2—
MY RESPONSE TO PD

MUSCLE BUILDING—JOB ONE

Parkinson's disease has several major symptoms that are characteristic of the condition—trembling (tremor), stiffness, slowness (bradykinesia), and difficulties with balance and posture. However every case is different, varying over time and changing with different conditions. About one in four PD patients do not develop tremors. Others may develop symptoms early in life, while still others experience a very slow progression of the disease. Indeed, my own case seemed to masquerade as arthritis and normal fatigue for over a decade before being diagnosed as PD.

I can now identify with several of the more common symptoms of Parkinson's disease, including poor posture, disturbances of sleep and appetite, lack of facial expression, muscular weakness, writing, walking, coordination, balance and speech.

That's a pretty big list of problems, with far-reaching consequences for a person's quality of life, which leads into a discussion of treatment options. I am taking a dopamine replacement drug plus a dopamine agonist drug (which stimulates dopamine receptors). These drugs do have a positive affect on my condition, but a drug therapy program coupled with regular exercise is even better. And while we are discussing treatment, one should get plenty of sleep at night, rest several times during the day, eat a proper diet, avoid smoking and alcohol consumption, and have a positive mental outlook.

While we cannot go to the heart of the PD problem and eradicate the cause, whatever that may be, we can attack the problem of muscular weakness. This effort should pay big dividends by improving the strength of the muscles, exercising joints, helping improve coordination, balance and other PD related problems. By my observation, and in my case, my muscles appear to respond to frequent vigorous exercise. Unfortunately the reverse also appears to be true: if I miss several workout sessions, I seem to lose a bit of strength and muscular endurance. But don't worry—keep working out, because Muscle Building is JOB ONE.

COORDINATION IS EVERYTHING

Work together or die alone. You must have seen or heard that axiom before. Good examples include a military unit, or a football team, where teamwork is essential, otherwise lives or games can be lost.

The human body is another incredible example of a virtually perfect well-coordinated system that is online 24 hours per day for as long as we are alive. Its sole purpose is to KEEP US ALIVE. The smallest activity in the body is the product of the body's coordination center.

To illustrate, suppose you are stung by a bee on your left hand. The toxin in the bee's venom begins to permeate the area of the bite. After a short delay, the network goes into action, having sensed chemical entities that do not belong in your body. It sends you a very personal message—we have been attacked by a bee. Thousands of years of human development have culminated in a warning signal in the form of pain. The coordination system works again and you seek treatment.

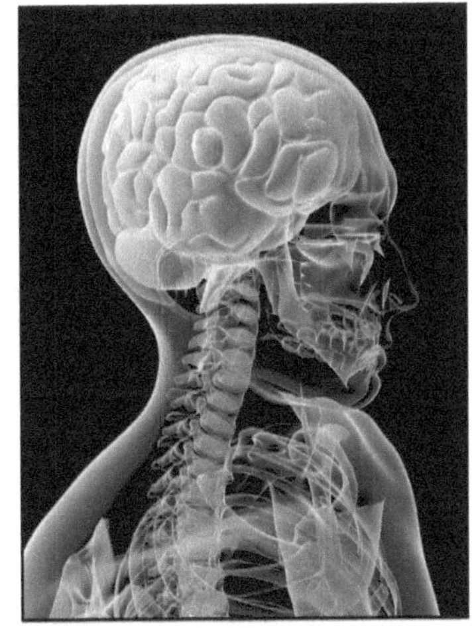

Every human body has a complex internal sensor network that tracks every condition, activity, abnormality, the good and the bad. A response is automatically generated, sometimes to correct a problem, sometimes to issue a new control signal, some voluntary—others involuntary, but in every case this adjustment is a coordinated action involving the entire body.

Many times we cannot observe the hidden activities in this network, but we can observe the face and its many expressions, the skin, and movement of the skeletal system and muscles.

Movement of the body in walking requires perfect coordination of the skeletal system, the muscles and the central nervous system. For safe walking, the body calls upon its organs, including eyes to look at terrain and obstacles, the inner ear for balance information, and the joints, skin and muscles. An incredible volume of information is processed by the brain, evaluated and transmitted via the central nervous system. And the subject walks, without even thinking about the mechanism that made it possible. In order to be able to walk, fifty-four different muscles in the feet, legs, hips and back must work together in perfect coordination.

In another example, to lift a hand, the shoulder has to be in a bent position, the front and rear arm muscles—called "triceps" and "biceps"—have to contract and relax, and the muscles between the elbow and wrist have to twist the wrist. In every part of this action, millions of receptors in the muscles transmit information immediately to the central nervous system describing the position of the muscles. In response, the central nervous system tells the muscles what to do in the next step. Of course the subject is not aware of any of these processes, but just wishes to lift his hand, and does it right away.

But when ***Parkinson's disease*** disrupts the human body, dopamine, an important chemical messenger in the brain, is in deficit. As a result, muscles become weak, the nerve pathways to and from the brain are diminished and the fine control of the body is disrupted. Coordination suffers. The PD patient can no longer walk smoothly, stand up as easily, or take care of himself as well. Medical treatment and a vigorous exercise program can mitigate the symptoms of PD, but realistically, life for the PD patient is changed by the disease.

SWING THOSE ARMS

Parkinson's disease affects muscles, joints and posture indirectly as a result of an undersupply of dopamine needed to maintain neurological command circuits in the brain. The muscles are not connected to central command in the way that they should be and therefore they become weak and inadequate in providing the range of functions needed throughout the body.

Reduced physical activity is a consequence of reduced muscle performance. For example, someone who played tennis before experiencing the limitations of PD might stop playing tennis. Then the shoulder, elbow and wrist joints that were used in tennis might become stiff. Unless these joints are exercised, they may acquire adhesions that limit the range of motion of the joints.

Joints have a certain amount of rotation allowed by the bone structure. There is a gap between the joints that permits them to slide over each other, facilitated by a lubricated sac. However, an inactive joint is susceptible to becoming immobilized and difficult to move. Its **Range of Motion** is the degree of movement allowed by the geometry of the joint.

It is important to keep exercising joints, especially shoulder joints. One relaxing way to exercise the shoulder joints is to take a hot shower and raise the arms as far up and back as possible. Then swing your arms down and back as far as possible.

DON'T MAKE A FACE

The human face displays **expressions** that convey important non-verbal communications. While the general appearance of a human being is fixed by the bone structure in the skull and face, the appearance of the face changes due to actions of over 40 muscles located beneath the skin in the face. "Smile" and "frown" are familiar forms of non-verbal communications that nevertheless say a lot, although silently.

Unlike other skeletal muscles which attach to bones, facial muscles attach to other muscles or to the skin. This means that a tiny contraction in one of the muscles in the face pulls the facial skin and changes the expression. By contracting facial muscles in different ways one can produce countless different expressions, from frowning to smiling and winking to raising an eyebrow.

The basic emotions that are associated with facial expressions include anger, sadness, fear, surprise, disgust, contempt and happiness. We tend to take for granted the important role played by facial expressions, until we lose them. Botox injections represent one way to interfere with facial expressions. Damage caused by Parkinson's disease also can nullify facial expressions.

Among the numerous facial muscles, there are several that do incredible things. For example, the muscle around the mouth controls the size of the mouth opening and is important for speech. It is also known as the "kissing" muscle because it is responsible for puckering the lips.

This muscle is very expressive and shapes the lips in different ways during speech. What would happen to a person's speech if this muscle stops working? Let's try an experiment. Say, "My dog has fleas." Now, say it again, but don't use your lips. See, it is almost impossible to enunciate clearly without using your lips.

Another important facial area involves the eyes. The eyes are surrounded by a set of muscles that open and close the eyes, and cause the eyelids to blink. If this set of muscles is weak, blinking may stop and the "windshield wipers" of the eyes (the eyelids) will not keep the eye clean. Then the eye can become dry, dirty and irritated or infected.

PD patients sometimes drool because the lip muscle is weak and relaxes instead of clamping the mouth shut. A stooped posture also can contribute to saliva accumulating at the lips. The lip muscle can be exercised by squeezing the lips tightly together, holding 10 seconds and releasing. Repeat 5 times. Another good exercise is to "pucker."

The Bottom Line: The best way to retain good facial expressions is to exercise facial muscles on a regular basis. Watch your face in a mirror while you smile, frown, and act surprised. Repeat this exercise 10 times.

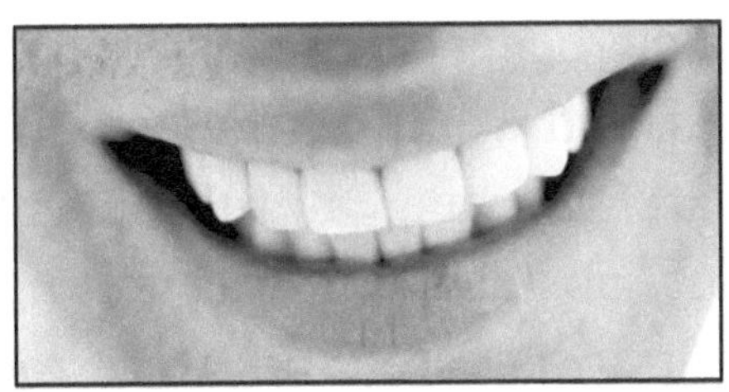

WALK TALL AND FAST

Did you ever see a "speed walker?" This odd looking walking technique is intended to burn off calories and get the heart rate up by walking very fast with a lot of arm motion. Its proponents claim that speed walking does not impact the joints as much as jogging, where both feet leave the ground at once.

When I commuted from New Jersey to mid-town Manhattan, I practiced my own version of speed walking, particularly in the late afternoon when the first express bus to New Jersey left the Port Authority Bus Terminal at 5:00 p.m. My office was about one mile north of the bus terminal, and I really had to move fast as I got off the elevator in my building at about 4:20 p.m., knowing that the bus passenger line was already forming. Thump-thump, I could feel the impact of my heels on the concrete sidewalk, which was loaded with people of all sizes and shapes, some moving fast, others moving slow.

Thump-thump— faster-faster—I commanded my legs. I was taking in air at a great rate as I reached Warp Speed. After dodging slow pedestrians, fast taxi cabs, kids with loud boom boxes, and would-be muggers (I was assaulted twice in 11 years) I swung at full speed into the bus terminal. The only obstacle now was getting up to the bus level via the stairs (the escalators were too slow.) Yikes, the bus is loading passengers, engine running, "Oh please," I thought, "wait for me!" The last three passengers for the Jersey-bound express bus were getting onto the bus, as the driver counted the number of people already on board. I looked him in the eye, and he nodded for me to get on the bus. I got the last seat. Thanks legs, we did it again. Hot, tired, out of breath, but triumphant. As I sat down in the back of the bus and looked at my watch (exactly 5 p.m.), I thought how good it was to win!

This little tale of walking tall and fast illustrates how much we rely on the wonderful bodies we have. All I needed to run this race every work day was a healthy body and the will power to do it.

Could I walk with that intensity today? Not likely, because Parkinson's disease has taken away some of my walking ability. Walking upright is extremely difficult and requires a lot of the human body's resources, including balance, coordination, muscle power, and flexibility just to take a few steps across a smooth floor in a straight line. But stepping out into an unfamiliar uneven terrain with many obstacles in the path requires even more leg power to adjust a step, alter foot placement, and sense ground resistance—things most people do without conscious thought.

I can sense that PD has diminished my sense of balance, interfered with my coordination in some way that I do not understand, weakened my muscles, and reduced my flexibility. As a result, my walking ability is less than normal. My movements are jerky, the muscles do not work together as well as before, and I must concentrate on not stumbling. I seem to walk better at a fast pace, possibly because the unevenness in my walk is shorter in duration.

Since walking is so important, we concentrate hard on exercises involving walking. I walk frequently around my neighborhood. I do side steps, and I march with knees high. These exercises have helped retain my ability to walk. I think that most people would not notice anything much wrong, but of course, these people can walk without thinking about it. I have to constantly think what I am doing. I suppose I'll never get my ease in walking completely back again.

I miss walking. I loved to go out for a walk on the beach. It was really nice to walk around an art exhibit without struggling to walk. Shopping at Christmas time was a wonderful treat—all those colorful displays, bright lights, Santa Claus in his fancy red suit and white beard. But no more Santa. I sometimes get a bit dizzy walking in an airport terminal building, maybe it's psychological. All I know is that walking will never be fun again, but I'll hang in there and enjoy as much as I can.

THE STRAIGHT STORY ON POSTURE

Stand up straight. Shoulders back. Chin in. Chest out. Eyes forward. Feet together. Are these just good suggestions? Not if you are a U.S. Marine Corps recruit. That Marine DI (Drill Instructor) is in his usual bad mood and he means business!

Well, why not follow these commands? I only wish that I could achieve the results the DI seeks now. When I was a raw ROTC Cadet in summer camp, I did a great job of military correctness on the drill field. My posture was "ram rod" straight. You would be proud of me. The DI approved too.

Due to Parkinson's disease, my posture has gotten worse over the last several years. The classic case of PD describes a progressive stooping of the body, so that the PD patient bends forward at increasing angles. I have been trying hard to stop this downward drift in posture by exercising the back muscles. I think the slide is not getting a lot worse.

There may be more to my posture problems, however, than just PD. Several years before the diagnosis of Parkinson's, I was diagnosed with a disorder of the spine with a long name, spondylolisthesis (spondy, for short). In short, my spine has ruptured discs and a displaced vertebra. Fortunately the vertebra is not touching the spinal cord. The orthopedic doctor who read an MRI of my lower spine remarked that he was surprised that I do not have more discomfort. While I am usually somewhat uncomfortable with minor pain from my back, and I have had sciatica, my condition is bearable.

Which came first—PD or spondy? Or does it matter? Well, I just would like to know whether PD has damaged my spine, or were my back problems going to happen anyway. It seems that a large number of people have back troubles—PD hardly could account for all of them.

Furthermore, I had a serious back problem before I was diagnosed with PD.

When I see my posture in a mirror, and when my personal trainer reminds me, I make a renewed effort to strengthen my back, abs and shoulder muscles. I realize that the only way to control stooping is to build these "core" muscles.

LIFE HANGS IN THE BALANCE

We sometimes take some very important functions in the human body for granted. Before I got Parkinson's disease, I had very good balance. Just what does that mean? There is a delicate mechanism in the inner ear that senses the position and motion of the body and sends balance information to the brain. The brain, in turn, develops signals to the muscles that control the body's position. The sole purpose is to keep the body from tipping over and falling due to the pulling force of gravity.

These mechanisms that control balance are so efficient and are so perfectly integrated with other parts of the body that normal, healthy people seldom fall. Injury or illness can interfere with the balance center in the inner ear. Some people have balance problems due to an ear infection, or a cold. These balance disturbances are usually reversible after the source of the problem is treated.

Then along came PD. The disturbance to my balance caused by PD is less pronounced and different in its nature than, for example, an inner ear infection. But my balance is not as good as before PD.

My Personal Trainer devotes a major effort to balance training because of the risk of a serious accident caused by my losing my balance and falling. We practice again and again, over and over. The good news is that balance can be improved with practice.

I am confident of my ability to maintain my balance as a result of constant balance training and practice. I can stand still in a waiting line, or walk a straight line between obstacles (for example, rows of seats) or walk up and downstairs without holding on. But be careful—don't fall, your safety hangs in the balance.

WHAT'S THAT?

I can usually tell when someone is having trouble understanding what I am saying. Their eyes cloud over, or they squint just a bit, or they turn their head just a tad—or they just come right out and ask, "What did you say?"

Persons with Parkinson's disease sometimes have trouble with their speech. In my case, my voice is softer, I am sometimes a bit hoarse, and my words are slightly slurred. The vocal chords are actually muscles which can degenerate due to PD, but they can be exercised like any other muscle in the body.

In every workout session with my Personal Trainer we perform voice and tongue exercises, which have helped me greatly to retain most of my speech. I actually have a strong, penetrating voice when I take a deep breath, bear down, open my mouth wide, enunciate clearly and project my voice. My voice is similar in volume and quality to the command tone I had in the military fifty years ago.

However I have a problem occasionally with slow swallowing that can interfere with my speech. This results in saliva collecting around my tongue. I sometimes have to stop talking briefly to swallow the excess saliva.

I am also still trying to train myself to use an appropriate level of voice control for the situation. For example, I may need to speak softly, but not whisper in church. Conversely, I may need to project a strong voice if I am giving a toast for a table seating 15 people. But just how strong? Even in a controlled and presumably predictable environment like my own kitchen, I have little confidence that my voice matches the needs of the moment. So I have resorted to the "can you hear me now?" chant heard in a cell phone network ad on TV. I suspect that my interpretation of the audio clues that I get from the room and my own voice are less reliable due to changes in my brain and central nervous system brought about by PD.

The good news is that with effort, most PD patients should be able to speak well enough to avoid the listener saying *"What's That?"*

DON'T GET COLD FEET

Did you ever shake hands with a person who had unusually cold hands? The frigid contact can be quite a surprise, causing one to pull away. I have some tendency for cold hands and feet, so when I read that cold extremities may be caused by Parkinson's disease, I decided to learn more.

It would come as no surprise if it turns out that PD, in some way, can affect the temperature of a person's hands and feet. However, I could not find any direct evidence on the Internet of PD's involvement.

Cold hands and feet are fairly common, usually caused by cold weather, not dressing properly or poor circulation. Most people warm up quickly. But in my case, it can take quite a while for my hands to warm up.

The medical literature reports numerous causes of cold hands and feet, including **Raynaud's** phenomenon, a condition where constrictions of the small blood vessels to the hands and feet interfere with normal blood circulation. These constrictions sometimes are caused by stress. Sometimes there may be a skin color indicator of the state of circulation to the hands—white skin, caused by lack of blood flow— blue to red skin, caused by a return of blood flow.

I think that it is safe to say that I won't get "cold feet" about cold extremities. Oh yes, a quick way to get your cold hands warm again—run warm water on them.

If we all did the things we are capable of,
we would astound ourselves.

Thomas Edison

A wise man adapts himself to circumstances as water shapes itself to the vessel that contains it.

Chinese Proverb

— 3 —
WHAT I LOST FOR GOOD

CARS DRIVE ME CRAZY

I have a passion for the automobile—that beautiful machine that once had a grand name like Buick. My father drove a **1927 Buick Sedan** when we lived in an Oklahoma oil boom town, where my father was superintendent of a small oil refinery. That's how my family happened to be living in Drumright. I was born there (in nearby Tulsa).

1927 Buick Sedan

As soon as I was big enough, I enjoyed sitting in the driver's seat of Dad's Buick sedan. Once in a while he let me steer while I sat in his lap.

1938 Pontiac Sedan

In the mid-1930's my family moved to New Jersey. My parents decided to buy a new car—I had wished for one of the new "streamlined" models without running boards, but to my disappointment they bought a **1938 Pontiac Sedan** with running boards. Just the same, I loved that old Pontiac like a brother.

World War II interrupted all manufacturing of civilian cars until after 1946, by which time our old Pontiac had seen better days. My parents finally decided to trade the Pontiac for a **1950 Buick Super Sedan** with the famous port holes in the hood.

1950 Buick Super Sedan

Off I went to MIT for the fall semester, car-less. I entered a contest in which the winner would receive a new Olds 88 Convertible, but I did not win the car.

After college I had two years of active duty in the Army Reserve. My first duty post was at Fort McLellan, Alabama, where I bought a new **1954 Chevy Bel-Air Sedan.**

1954 Chevrolet Bel-Air Sedan

My own wheels at last!

Cars are important to me. But my doctor said the unthinkable—I SHOULD NOT DRIVE ANY MORE—because my medication can cause me to fall asleep without warning.

"Do you know what this means?" I asked Jeanne later. "I'm going to be trapped in the house without a car. This damned PD is a life sentence in solitary."

She responded by saying, "But aren't you lucky that I can take you places? Remember, you are not alone."

A LOST TOUCH

My hands were my pride and joy. While I did not have big strong hands, I had something even better than brute force hands. I had the "touch." I could sense delicate details in an object by merely holding it in my hands and running my fingers over its surface. And I could hold delicate tools, like jeweler's files, and shape an object accurately. I could paint lettering onto an object, using very fine brushes and paints. These talents were very special to me.

So naturally I applied these abilities to projects that could benefit from a fine touch. I may have inherited a bit of my sister's considerable talent for art. As a teenager, I loved to paint water color pictures. I also made model railroad engines and rolling stock for HO model trains. I still have a handmade dining car complete with tiny place settings on miniature table cloths. I painted the railroad name on the side of the dining car using a micro brush.

My family and friends saw that I had a talent for fixing broken objects, like watch bands and the like. That led to frequent requests to fix this or that. I gladly obliged, because I liked to be useful and I was proud of my "touch." We took this talent for granted.

Hands like mine could draw and print like a draftsman. When I was in college, I could almost always get a better grade by using the "touch" to enhance the report, drawing or sketch. My accurate handwriting also showed in letters that I wrote. My engineering thesis was filled with beautiful pen and ink sketches of the laboratory equipment. My thesis advisor was impressed.

But it gets even better. I took piano lessons for several years starting when I was about seven years old. Boy, did I love playing the piano. My friends loved to listen too. I was pretty good at playing classical and popular music. My hands were not especially big, so I had trouble reaching some chords, but they were expressive — the "touch" at work again.

When I first went to work as an engineer, I used my manual dexterity to prepare engineering drawings that were well above average (there were no computers yet, so charts and drawings usually were hand-drawn). It didn't hurt my reputation for quality work — if Mike did it, it must be good.

Another aspect of my special ability for close-up work was my vision. I could see near-field objects very clearly without any optical aid, such as eyeglasses or magnifying glasses. This visual sharpness was a great aid to my hand-work, because I could immediately see my work without fumbling with an optical aid. While I accepted my sharp near-vision as a given—I did need glasses for distance. This near-sighted condition did not bother me—I just took off my eyeglasses for close-up work.

Visual perception and hand-eye coordination also are important for close-up work. To illustrate, in hand lettering it is important to dot the letter "I" and cross the letter "T" accurately. I seemed to do that with ease. I think my "touch" skills were a special gift. Perhaps I would have been a good surgeon or dentist. But in fact, I was fortunate that I did not undertake surgery or dentistry, because I completely lost my manual dexterity and visual skills by the age of 50. This loss of skill did not occur all at once, but gradually over a period of several years. Now I am certain that I was experiencing the early stages of PD, but at the time I just gradually lost interest in all of the fun things I loved to work on. No more watercolors, no more piano, no more model railroads, no more drawings or pen and ink sketches. People still asked me for help in fixing some gadget or replacing a tiny battery in a watch, and I would try, but I just could not do it anymore. I could not see clearly, I could not judge the size and shape of objects well enough, and I could not use tools accurately—in short, my "touch" was *gone for good.* I felt like I had two blobs for hands.

I had entered a new era. My hands did not work right anymore. I was "all thumbs." I could not see clearly even with glasses. I had to use a big magnifying glass to make out small type, which I used to do with no optical aid. My vision was distorted. I saw small "jaggies" along the edges of objects. My depth perception was off.

Unfortunately, as time went by, my hands got worse. It seemed that I could not predict what they would do when I attempted to use them to do very routine things. For example, suppose I need to copy the phone number 555-1234 onto a notepad. OK, I pick up a pencil, and then jot down the number—whoops—the number got scrambled to 555-2143. My brain, nerves, hands or something does not work right. I concentrate harder, still scrambled. The same thing happens when I try to type on a computer keyboard. My handwriting, which was so precise, has become hurried and scribbly. *Gone for good.*

PD PATIENTS NEED A BETTER MOUSE

Personal computers are really great. From their inception in the late 1970's to today's ultra powerful machines, the growth of power, speed, complexity, ease of use, communications, and Internet service is breathtaking.

This book was written on a modern notebook computer that weighs only a few pounds and can be stored in a small carrying case. The computer takes care of all the routine tasks, page numbers, headings, margins, spell checking, grammar, pictures, formatting, and much more. That leaves the important tasks, like thinking and writing, to the author.

Unfortunately, while computers have improved, my ability to use a computer has declined due to the progression of Parkinson's disease. The major problem areas are:

- Poor posture
    - The stooped condition of my posture strains my back muscles, which become fatigued and painful.
    - I spend much time and effort to try to ease

the pain — switch chairs, raise/lower PC, etc.
  - When no longer bearable, I stop to rest.
- Lack of manual dexterity
  - I can no longer touch type; I make many mistakes. My fingers are stiff and lock up.
  - I have tried several different computer mouse designs, but I have not found one that works well. My fingers sometimes accidentally touch the mouse buttons. A track ball integrated with a keyboard works well until the track ball becomes dirty.
  - When trying to move the mouse, my hand and arm tend to freeze. I sometimes move the mouse with two hands.

Overall, the computer makes writing a book much easier. The work of keying in the text, laying out the book, processing graphics, editing and proofing is simplified by the PC.

But if I could only feel and function like I did ten years ago, I could write a book faster and better. Nevertheless, the modern PC is a joy to use even with interference from Parkinson's disease.

It is hard to fail, but it is worse
never to have tried to succeed.

Theodore Roosevelt

Dost thou love life?
Then do not squander time,
for that is the stuff life is made of.

Benjamin Franklin

If you would know the road ahead,
ask someone who has traveled it.

Chinese Proverb

NOTE FROM THE AUTHOR

As I reviewed Part One, the preceding section, I thought at first, well—that should about do it—I told how PD has changed my life. Then I got to thinking, but how long did it take to get a good diagnosis? What problems did we encounter in finding a PD specialist? Did my symptoms improve with drug therapy right away, or if not, how much later? How bad were the drug side effects? Did my exercise program prove effective—what improvement did I see due to working out? And so on.

I realized that there was more to my story, because Part One of my book leaves too many unanswered questions. So I added Part Two, which covers the flow of major events over a period of several years, from my early PD years to the present.

If you don't know where you are going,
you'll end up someplace else.

Yogi Berra

Part Two

My Story

It is good to learn what to avoid
by studying the misfortunes of others.

Publius Syrius

— 4 —
STORM CLOUDS GATHERING

EARLY SYMPTOMS

So there I was, about 50 years old, working for a big company in New York City, planning early retirement and seeing Yankee games once in awhile. I seemed to be reasonably normal and healthy, even if a bit tired from a three hour round trip commute every work day. But all in all, things seemed to be going quite well.

Annual physicals indicated that I was in excellent physical health. You might ask why the examining physician failed to spot the PD sooner. Well, there is no specific test for PD. Its presence is inferred from symptoms. I did have a few symptoms, but I was not aware that these clues were important at the time. I never told the doctor - it seemed unimportant.

I understand that PD is caused by the failure of a small area deep within the brain to produce enough of a chemical called dopamine. This chemical serves a vital role in the functioning of the entire body. It is a chemical transmitter, connecting the control circuits in the brain to the nerve pathways to the muscles and, I suppose, to other bits of the body's machinery. I learned that two serious muscular problems occur in the absence of the necessary level of dopamine. First, the muscles can waste away and become weak. Second, the commands from the brain to use the already-weakened muscles are themselves weak, because an inadequate number of nerve cells to the muscles are fired. The combined result can be a near-collapse of the muscle system. There are over 630 skeletal muscles in the human body. Because of PD there probably is a ballet dancer who can't dance, and a surgeon who can't operate. Their bodies just don't work right. What a waste.

The deficiency in dopamine production in the brain is irreversible. The experts say this region of the brain becomes inactive. There is no warning. It just stops working, in my case, slowly. What is the cause of this failure? The researchers don't really know. Some think that the brain is injured by a virus or some organism, others blame pollution. Who knows? I had a couple of really bad illnesses that required hospitalization. Maybe that was the cause of my PD. I also worked in an oil refinery and chemical plant. Could that be the cause? Who knows? It does not matter now.

I noticed a change that I did not understand. I used to have excellent manual dexterity but around the age of 50, I noticed that my handwriting was becoming more difficult for me. I began to notice some loss of control of my right hand when writing. My writing got smaller, scribble-like and sometimes mixed up. Frequently my written words got out of order. These glitches were entirely involuntary. I thought it was just that I was older, or tired, or whatever. At work I often had trouble writing a short note, like "I can attend the meeting at 2 p.m." That simple note might take two or three tries to get it right.

A similar problem occurred with my use of a computer keyboard. I once could touch type at about 80 words per minute. My left and right hands can no longer do that. Same problem, scrambled up words, mistakes—well, I'm having the problem right now as I type. Very frustrating. Tiring. Slow - slow - slow. But I didn't tell anyone. I just said to myself, get over it. Do what it takes, you have an important job to do. Your family is depending on you. Hang in there.

Strangers notice my slowness too. For example, an appliance repairman asked me to fill out a form. I was slow in doing that, too slow for him, so he became impatient and finished the form himself. I did not report this slowness to my doctor.

It never crossed my mind that I could have a serious disease. I now believe that this hand writing problem was the first of several symptoms of PD. If that assumption is correct, I have been experiencing symptoms of PD for as long as 25 years. This could be good news, because I may have the type of PD that progresses slowly. However, this slow progression and subtle change made the onset of PD hard to detect. There never was an acute pain, or catastrophic event to alert me to trouble. No flashing lights. No blaring sirens. No alarm bells.

In 1986 I retired early at the age of 55, and we moved from our home in New Jersey to Florida, first to Venice on the west coast of Florida, then back to New Jersey for several years (Jeanne was not ready to retire yet), later to Delray Beach on the east coast and finally to Jupiter in Palm Beach County. We enjoyed Florida living: no more cold weather, no more snow to shovel, no more leaves to rake in the fall, no more long commutes. Neither Jeanne nor I were accustomed to having spare time after a life of work, work, and more work.

What do some people do with spare time? Why what else — play golf. So off we went to the golf stores, finally buying two sets of golf clubs. We had never before played golf, so we took lessons from a golf instructor. Jeanne did well and learned enough to begin playing golf with lady friends. However, I was another story. I absolutely could not hit the ball. The instructor was very patient, but he clearly was getting frustrated. I tried so hard that my wrists ached from the uncontrolled, increasingly desperate swings. PD strikes again. My clubs ended up in a closet until we sold them.

There were other symptoms too. I seemed to have less muscular strength. It was becoming more difficult to lift heavy objects. But my way to rationalize this decline in capability was that I was getting old. I also noticed that my "arthritis" hurt more, and my posture was slipping. Then one day I saw myself in a photo in which I seemed to be shorter than before. I had Jeanne measure my height, and found that I had lost a couple of inches of height. But don't tell the doctor. It's just old age; everyone loses a bit of stature as they age - or so I said. A stooped posture is common in PD, because the muscles that support the frame get weak.

My family physician advised me to get some cardiovascular exercise. We decided that I should walk in the neighborhood. I tried to keep up with Jeanne when we walked, but I got tired very quickly and had to give it up. I was getting discouraged, but I did not consider the possibility that I had a serious disease.

There was always a good reason to dismiss these symptoms as unimportant. I have always been slow to declare an emergency unless the need was apparent. The basic problem with my case of Parkinson's disease may have been that the onset and progression was very slow, and the weakness caused by PD resembles other conditions.

Still more symptoms emerged. I was getting stiff and less coordinated. The inability to walk properly is a symptom of PD.

Something was wrong with me, although I did not want to admit it. Jeanne had noticed my physical problems too, and she concluded that I needed to exercise more. Therefore, at Jeanne's urging in early 2004, I joined the Wellness Center at the Jupiter Medical Center in Jupiter, where I worked with Ms. Maribel Bleeker, a Personal Trainer assigned by the Wellness Center. For the first time since rowing on the crew at MIT, I undertook a disciplined workout program. But I had trouble. I was weak and uncoordinated. I found walking on a treadmill or riding an exercise bike to be exhausting. I could not go for more than a few minutes at a time on any machine. I didn't seem to be making much progress in the gym, which was puzzling to the trainer. She was very patient, but I was unable to improve my performance in spite of her efforts. I tried to get my heart rate over 100 on a treadmill, but I just could not do it. I also had trouble maintaining my balance.

Then one day my personal trainer said, “I don’t want to discourage you, but you have spent the last several months trying to raise your performance level. To be honest, I think you are at a standstill.”

I replied, “I guess you’re right. I have been thinking that I should have a physical exam.”

“That’s a good idea,” she said. “See what Jeanne thinks.”

Jeanne agreed that the time had come to get our doctor’s evaluation. So the first thing I did after we got home was to make an appointment to see our doctor.

— 5 — BAD NEWS

A DIAGNOSIS — PARKINSON'S DISEASE

Don't worry, I thought, as I parked at the doctor's office. I had an appointment in a few minutes. I checked in at the receptionist's desk and took a seat in the waiting room. I have had many physicals; this one is no different. There must be a simple explanation for my weakness and other unexplained symptoms. Maybe I need some vitamins, more sleep, better workout clothes or shoes. I was not sure what to expect. I was a bit uneasy but I never thought that I might have a serious disease. Therefore, I was not well prepared for events as they unfolded.

I flipped through an old magazine in the doctor's waiting room. The nurse opened a door into the waiting room and said, "Michael Stanfield?" in a half-questioning tone of voice. Boy. They had bad chairs in the waiting room—I had trouble getting out of the chair (eventually I learned that this was another symptom of PD). In a few minutes the doctor joined me in the examining room. I explained why I was there. She listened intently, and then she asked me to take off my shoes and socks and walk around the room. Next she asked me to close my eyes and stand tall. After that she had me reach out and touch her finger tips with my index finger. She was silent. Her expression was difficult to read. I suddenly felt uneasy and I sensed that the color was draining from my face.

Then she spoke quietly, as if not to upset me. "Michael, I would like for you to see a specialist, a neurologist." I asked what for? She said she couldn't be certain, but I might have symptoms of a neurological nature. For the first time I got concerned. I thanked her and got the referral information as I left her office. What was this all about? I went home and told Jeanne. Now we could both worry—brain tumor—Alzheimer's—stroke—old age?

We called the neurologist's office and got an appointment on the following Wednesday at 11:30 a.m. I cannot forget the date—June 9, 2004. Oh great, now the worry wart part of me really had something to worry about. The big day came. Off I went to the neurologist's office in Palm Beach Gardens. This doctor was really busy. The waiting room was overflowing with people who were also tired of waiting. I waited for almost an hour until I was finally escorted into an examining room. It was a tiny room, not much bigger than a closet.

The door opened and the neurologist, a younger woman, entered. She seemed to be in a hurry when she asked why I was there. I nervously replied that my family doctor had referred me. She had me perform several tests involving walking and movement of my hands and arms. Then she sat down, looked me straight in the eye, up close and face to face, and in a curt tone of voice said,

"Mr. Stanfield, YOU ARE SLOW! Your eyes don't blink and you don't smile." She paused briefly and then went on, ***"You have Parkinson's disease."***

Bang! I thought that I had been shot. I sat there stunned as she went on to tell me that some patients see an improvement when they take an over-the-counter "dietary supplement" called Co-Q-10. She recommended that I take 600 milligrams daily and then come back to see her in about a month—good-bye, have a nice day. I was dumbfounded. Was *that* it? *Come back in a month???* The neurologist got up and left me to ponder what she said. Frankly, I don't think she cared if I came back.

I staggered out of her office and drove back to Jupiter; my head was spinning. My first stop was the drug store, where I loaded up on Co-Q-10, which cost almost one hundred dollars (expensive stuff). Jeanne was not home, so I quickly gulped down the full daily dose and tried to reach her on her cell phone. No answer. While waiting, I read the label on the Co-Q-10. It is supposed to improve muscular health by reducing harmful "free radicals." Side effects include nausea and dizziness. They got that part right! I was terribly nauseous. I was miserable by the time Jeanne arrived home. I told her my story. She was annoyed that the neurologist had been so insensitive in her treatment of me. Jeanne said that she should have gone with me. She also said that I should not have taken 600 milligrams at once.

I guess I had lived a very routine and sheltered life. I never before had to face the grim reality that our lives would be impacted by a major illness. But that day, as I sat and thought about the negative things the neurologist said, I was devastated, frightened and depressed. It seemed like a death sentence, without hope. I could not stop thinking about my becoming like the late Pope John Paul II, whose advanced symptoms of Parkinson's disease were apparent during public appearances. I would rather die. We really didn't know much about PD so Jeanne went to the Jupiter Library and checked out several books on the subject. She read some passages to me, but while interesting, these books also added to my anxieties.

I had a really bad night Wednesday. I could not sleep and my restlessness kept Jeanne awake too. My mental state was even worse on Thursday morning. Jeanne went into action. She called the offices of our family physician and the neurologist and asked for a prescription for a tranquilizer from one doctor and a sleeping pill from the other. "Please call it in," she asked. By 2 p.m. nothing was heard from the pharmacy, so she again called both doctor's offices to inquire about the status of the requested prescriptions. This time she was told that the doctor didn't like to prescribe drugs containing tranquilizers. She explained that I had gotten a deeply upsetting diagnosis and needed something to sleep for a few nights. Around 4 p.m. she got a call from the pharmacy that my prescription was ready. Well, it was about time!

When Jeanne got there, the pharmacist held up two bags, both containing a prescription, one for a tranquilizer and the other for a sleeping pill, which he said were essentially the same thing. He said that I could not have both. Jeanne said that she was a registered nurse, and she would not allow me to take anything harmful. She told the pharmacist that she needed the flexibility of having both medications available. He agreed reluctantly. She had spent the whole day getting a couple of simple prescriptions filled.

It was late afternoon. I took the tranquilizer, which did help me to relax a bit. She read more from the library books. One of the books said that PD has five stages. Jeanne thought that my symptoms put me in stage three. She said that she would not tell me about stage five, it might be too upsetting.

I took one of the sleeping pills at bedtime. It did put me to sleep, sort of. I think I was in La-La Land the whole night. It was unreal. Apparently I didn't respond well to this kind of sleeping pill, whatever it was. I woke up in the middle of the night with a half awake, half asleep feeling. The world seemed to be revolving around my head.

But where was I? I woke up gradually, slowly sensing that I was in strange place. ***It was a strange place!*** I could see parapets with gold spires surrounded by yellow sand. In the background was a rusty orange sky. Several men in Arabian garments walked slowly along a sandy path, as Middle Eastern music sounded softly. Do they see me? Oh boy! One of the men had seen me and was running toward me. "I must hide," I thought. So I dove headfirst into a pit and felt myself falling. Bam, I hit bottom. That woke me up for sure. I had fallen out of bed and gotten stuck between the bed and the nightstand. Man! How was

I going to get out of there? Was this a dream? No, it was really happening. I struggled, pushed, pulled, squirmed, and finally got free. Wow, what an ordeal!

Somehow I pulled myself back onto the bed, but I was lying *across* the bed. Jeanne said she woke up with a heavy weight on top of her. It was me, continuing my Night in La-La Land. She pushed me off of her but now she couldn't sleep anymore, so she got up. I finally zonked out for good that night.

This comedy routine came to an abrupt end on Friday morning when Jeanne addressed me in an all-business tone of voice: "OK, buddy. You have had two days to feel sorry for yourself. Now the pity party is over. Get up and get dressed. We have work to do! We have got to get you a good specialist in Parkinson's disease. Let's GO."

“Yes, Ma'am. Understood, Ma'am.” As an Army veteran, I knew an order when I heard it.

We ate breakfast and Jeanne cleaned up the dishes while I looked at a PD book. I had begun to get back my sense of humor as I joked about my sleeping pill episode.

Jeanne has an unusual ability to communicate with a few individuals by some sort of mental telepathy. She said later on that she sent a mental message to Mike that she needed him. The phone rang. It was Mike on his cell phone, while on the way to work. "Ma—is everything OK?" he asked. She told him what the doctor said, and that we needed to get a good PD specialist. He stopped his car, turned around and went back home. He told her that he would call back.

Mike was at work when he called back again. Mike is

employed by the New Orleans Saints NFL football team as Vice President of Ticket and Suite Sales. He had spoken with the Saints owner, whose late wife had PD, and got the name of the specialist who had treated her. Mike said that a Dr. Rao would call her in the next ten minutes or so.

"*Yeah, sure,*" I said sarcastically. "*He's a specialist and yet he will call us?* ***I bet****. Not likely.*"

A few minutes later the phone rang. The Caller ID said "out of area." We thought it was just another unsolicited sales call. (We get nuisance calls frequently.) Then the answering machine picked up and a friendly voice was heard to say, "Hello. This is Dr. Rao at the LSU Healthcare Clinic in New Orleans." Jeanne immediately picked up the phone and thanked him for calling us. He wanted to know about my case. Jeanne summarized the recent diagnosis by the neurologist in Florida. He commented that our son Mike had said that I was in Stage Three of Parkinson's. Jeanne responded by admitting the Stage Three description was her own estimate as a former registered nurse. Dr. Rao shot back, "Even if you are a nurse, you should not be diagnosing, please leave that to me."

"What medication is he taking for the Parkinson's?" he asked. When she told him that I was not taking any medication, he practically exploded. "What! No medication? He needs to be on a treatment program immediately." Now, *that* was music to my ears!

He then asked whether we could come to New Orleans right away for an examination. He said that he wanted to be sure that my problem is really PD, because there is another disorder with similar symptoms that is much more serious than PD. (This disease is fatal.) Jeanne said that we would be on a plane to New Orleans Monday. He replied, "Good. I'll see Mike's Dad at 9 a.m. Tuesday morning at the LSU Healthcare Clinic on Gravier Street. Your son Mike will have all the details by the time you get here." We were overjoyed that Mike had been able to find Dr. Rao on such short notice. We spent the rest of the day making travel plans and packing for our trip to New Orleans.

Boy! Was I lucky! This doctor is a world class specialist in PD, and yet he is willing to see me in a couple of days! Taking positive steps is infinitely better than sitting around and feeling sorry for oneself.

I thought to myself . . .

I'm going to beat this disease!

— 6 —
FIND A SPECIALIST

DR. RAO TO THE RESCUE

I was nervous and apprehensive as Jeanne and I boarded a Continental flight scheduled to depart Palm Beach International Airport at 7:00 a.m. on Monday. The plane was filled with what appeared to be business passengers on their way to their next sales call. These passengers seemed unconcerned and relaxed as they settled back to read their newspapers and sip their coffee. I was anything but relaxed as the plane became airborne. I fidgeted in my seat. The plane banked sharply to avoid flying over residential areas of West Palm Beach while it rose higher and higher into the clear blue sky.

We were on our way! I felt somewhat encouraged that I was to be examined by a specialist, but I had much anxiety about my long term outlook. It was difficult to get the image of the late Pope's affliction out of my mind. As I pondered my fate, I nibbled on a snack served on the plane. Jeanne had only a cup of coffee. Neither of us was especially hungry. She was really tired after a couple of frantic days getting ready.

None of the airlines offered non-stop flights from West Palm Beach to New Orleans. Delta had a route that flew into Atlanta, then after a change of planes, on to New Orleans. Continental's route required a change of planes in Houston from a Boeing 737 to a much smaller Embrauer commuter plane. We disliked the commuter flights because the planes were small and uncomfortable, and they seemed less safe than the bigger planes. Another inconvenience was that full-sized aircraft, like the 737, used Continental gates that were far from the gates assigned to Continental commuter flights. In fact, the regular gates and commuter gates were about as far apart as they could be. This meant that we had a long walk to get the connecting flight, and if the first flight was late arriving in Houston, we had to make a mad dash to avoid missing the next plane. Try that if you have PD - it's not fun.

We arrived in New Orleans in late morning, and we were met by Stacey who escorted us to an arriving passenger area where Mike was circling in his SUV. From there we traveled a short distance to their home in Kenner, where we stayed until our return to Florida. Kenner is northwest of the city of New Orleans. Mike's job with the Saints took him to Metaire, where the Saints business offices and practice fields are located. At that time, Stacey was employed in marketing by Six Flags Amusement Parks.

Her territory included the City of New Orleans, so she was familiar with the location of the LSU clinic where Dr. Rao was located.

We were greeted by Simba, the ever lively puppy, who was excited to see us. Simba is a Cavalier King Charles Spaniel, a small but playful breed. Simba liked to play "catch me if you can" with socks, slippers, pieces of paper, anything within his reach. In spite of his diminutive size, he could jump up almost into your lap while you were seated at the dinner table to snatch your dinner napkin. Then, off to races and "catch him if you can." If you happened to corner him, he would sink his teeth into whatever he took so hard that it was impossible to get it back.

Mike went back to work, while Stacey fixed us some lunch. We were very grateful to be able to stay in their home, rather than a downtown hotel. Stacey kept us well-fed during our stay. She fixed great breakfasts, lunches and dinners. Our stay was like a vacation, except for one minor detail—I had a problem called Parkinson's disease.

This brings me to the events of Tuesday morning. Stacey also was our driver and sightseeing guide in our trip downtown to see Dr. Rao. We took I-10 into the City, which was loaded with rush hour traffic. The road patterns in downtown New Orleans seemed to be complicated, but somehow she negotiated all the numerous turns, one way streets, traffic lights and narrow streets and dropped us at the LSU Healthcare Clinic on Gravier Street. We went to the Fourth Floor,

Neurology, where the receptionist took my insurance information and then asked us to wait in Section C.

The Clinic was huge, much bigger than we had expected. The waiting room in Section C was packed with patients and family members. This scene was an unsettling eye-opener. There were elderly men with expressionless faces, some with tremors (uncontrollable shaking), some in wheelchairs, apparently unable to walk. There also were very young adults, accompanied by parents, some with their head flopped over onto their shoulder, others with tremors. This was quite sobering for me. A sign posted on the wall read "Section C, Movement Disorders." I seemed to be in better shape than most of the other patients. The world was filled with so many people in misery. Again I thought how lucky I was to see Dr. Rao only six days after the diagnosis. With that, a nurse called my name, and we entered the patient examining area, where the first procedure was the measurement of my height and weight. She also took my pulse and blood pressure. I was really uptight, and my blood pressure, usually normal, was high that day. Then Jeanne and I waited for Dr. Rao in a small examination room.

The door opened and in came Dr. Rao. He was a slender man, about 5 ft.-10 in. or so tall, with a big grin, and a hearty hand shake. His full name is Jayaraman Rao. I said, "Your reputation precedes you, Dr. Rao." "Not all bad, I hope," he replied with big smile. He told us about the poor victims of Parkinson's disease in his native country, India. They lay in hospital beds, where the only mercy available was to turn them occasionally. But that is not what he had in mind now.

First, he examined me by taking hold of my wrists, one at a time, lifting and lowering them. He also took hold of my arms, raising them over my head. He asked a few questions. Then he turned to Jeanne with a big smile and said, "I apologize. You are right, he is in Stage Three. Well, we are going to do something about that." I liked his positive attitude. I asked him if I would end up in a wheelchair. "*No, no, no.*" he replied. "With treatment, you will be fine!"

Dr. Jayaraman Rao, MD

This initial part of the exam took less than fifteen minutes. Dr. Rao looked at me and said, "I am going to ask you to take this medicine, and go back to the waiting room. I'll check on you in a few minutes. Please note how you feel." I did as he asked. In a few minutes, Dr. Rao's assistant, whom we met briefly on our way in, came out and asked me if I felt any different. I replied that the room lights seemed much brighter. She said the doctor would see me shortly. Indeed, he reappeared in about twenty minutes. Dr. Rao again flexed my wrists and arms. I sensed that my response to the pill was not as much as he had hoped for.

But I did respond! My eyelids, which usually were drooping, were wide open! He had given me the magic pill, carbidopa/levodopa, which replaced some of the missing dopamine in my brain. While limited, the response gave me encouragement.

Then he popped the question. He asked if we could stay for a week so he could monitor my response to a daily intake of the medication. He said he would call Mike to see if that was OK by him. He did call Mike, and Mike said yes. So our tour was extended by mutual agreement to Friday, not quite what Dr. Rao had wanted but it would have to do.

We returned to see Dr. Rao late in the week. He said he had good news and bad news. The good news - I had responded nicely to the PD medication - the bad news - I could not have it yet. He explained that we needed to save the "Good Stuff" for when I really need it. For now he wanted me to take a drug called **Requip**, for which he provided a dosage kit. This kit contained several different strengths of medication, starting with a light dose and getting progressively stronger. This drug can cause severe drowsiness, so Dr. Rao advised against my driving. He also explained that Requip is a dopamine agonist, which helps to protect the brain and assists whatever dopamine is still being produced by the brain.

Requip helps relieve the symptoms of Parkinson's disease, which is caused by a deficit of dopamine (one of the brain's chief chemical messengers). Requip works by stimulating dopamine receptors in the brain.

One final issue needed to be resolved in an otherwise encouraging visit. I was concerned that Dr. Rao was so far away in New Orleans that visits would be hard to schedule and expensive. Dr. Rao suggested that we would be in New Orleans anyway to visit Mike and Stacey. He said he would see us anytime we were in town. But I thought we needed to have a PD specialist closer to home. The good doctor and his assistant went online and suggested a qualified neurologist at Cleveland Clinic in Weston, Florida (near Fort Lauderdale).

So off we went, back to Jupiter, after thanking Mike and Stacey for their hospitality and their assistance in finding Dr. Rao and taking us to see the doctor. Jeanne and I both had renewed optimism.

I began to experiment with different doses of Requip, as we arranged to meet the specialist in the Cleveland Clinic. The long road to improvement still lay ahead, but at least we had begun the journey.

The razzle dazzle of our trip to see Dr. Rao, which had pumped up our morale, was now replaced with the daily routine to determine a suitable dose for Requip.

Just what were my symptoms in the summer of 2004? That would help structure my activities by identifying things I should NOT do. It was easy to make a list of symptoms. I sometimes fell asleep without warning. I got tired easily. I lacked muscular strength. I was uncoordinated. I had difficulty walking. My manual dexterity was poor. My handwriting was small and almost illegible. My balance was poor. I had trouble getting out of a chair. I had a problem swallowing and drinking from a glass. My speech was hard to understand and I mumbled. I sometimes drooled without warning. I was very slow. My perception of distance was distorted. I tended to place objects too close to the edge of a table.

Referring to the range of symptoms above, here is a list of activities that were deemed to be unsafe.

DON'T

- *Drive a car*
- *Operate dangerous machinery, like power saws*
- *Climb on ladders*
- *Carry heavy objects (like furniture)*
- *Walk where there are tripping hazards*
- *Work on dangerous repairs, like electrical wiring*

But I could still do some things that were reasonably safe, useful and interesting. I had to be productive and busy, or I would run the risk of a mental decline too. As previously mentioned, I have an interest in music and writing. If only I could get my symptoms under control I would be very productive.

One of the most disruptive aspects of PD I found to be the slow decline in physical strength, flexibility, coordination and balance. This multi-pronged attack on the human body can be devastating. How could I work on music if my body was collapsing around me?

We know from experience that exercise is a vital part of treatment for many cases of Parkinson's disease. As Jeanne likes to point out, she urged me many times to have a regular exercise program. Furthermore, my employer had a gym in its New York headquarters. I was eligible to use this gym during working hours at no charge. That was a wonderful opportunity — of course, I never used it. Too busy, I said.

Jeanne was right. Now I have a serious reason to exercise hard, it's called Parkinson's disease.

I have a dream: I want to be a "real person," who can walk, talk and think like anyone else. But to do that I need a functioning body.

How do I achieve this dream? Medication and exercise is the two-part answer. The medication part helps to supply some dopamine to make the body operable. The exercise part addresses the vast mechanical complex of the human body, **the muscles**, which operate the human body.

I am committed to a regular program of exercise, supervised in part by a knowledgeable personal trainer. I go to the gym for cardiovascular exercise, and strength building. I also have worked with a Speech Therapist as an outpatient at the Jupiter Medical Center.

My personal trainer, Ms. Maribel Bleeker, a Health Fitness Specialist, devotes an hour three times a week to conducting an exercise program for me in my home. This program is described in Appendix I (page 132). See Appendix IV (page 142) for Maribel's qualifications, training and experience.

In the summer of 2004 I used the dosage kit provided by Dr. Rao to determine an appropriate strength of Requip. Every week I opened another container of tablets, each one a small step stronger, as measured in milligrams. The dosage differences from week to week were quite small, but we finally decided that 3 milligrams was about right. This translated to nine milligrams a day, taken in three separate doses of 3 mg. each. I got a prescription and began the regular program of medication. We were also guided by Dr. Rao's experience with Requip that suggested around 9 to 12 mg. per day should be about right. Dr. Rao's office called in a prescription for a six month supply.

Requip has a few side effects, including drowsiness, upset stomach and nausea. I experienced most of these side effects, which lasted for half an hour or more. I had to remind myself that these feelings were just transitory, and would pass in under an hour. I was unable to identify the exact benefits of Requip, except that I seemed more energetic and better able to cope with daily activities. However, on the downside I sensed a depressed feeling after each dose of Requip.

I continued to be free of tremors and freezing (the temporary inability to move, an embarrassing problem should it occur in public places).

CLEVELAND CLINIC IN FLORIDA

In July 2004 we made an appointment to see the neurologist at Cleveland Clinic in Weston who was recommended by Dr. Rao. Weston is a community west of Fort Lauderdale, close to Alligator Alley, I-75, which runs east/west across the Florida peninsula toward Naples. We relied on the GPS navigation system in our car, which took us south from Jupiter on I-95 to 595 West to I-75 South, exiting at Weston. It was a nightmare drive that took almost an hour and a half in unbelievably heavy traffic. This drive was extremely stressful for Jeanne, who was worried about getting back home in rush hour traffic.

The Cleveland Clinic medical complex was huge, including a hospital and a separate multistory building for the clinic. We checked in at a large reception room on the first floor, and then took seats in the patient waiting room. Finally a nurse called my name and we were escorted to the examination area. In a few minutes a pleasant young woman doctor introduced herself as the neurologist's assistant. She took down some information about my PD. I asked her whether excessive saliva and a runny nose that I was experiencing were caused by PD. She replied that the neurologist would cover that.

Jeanne and I waited for the PD specialist in a small examining room. The walls were covered with posters informing the patient about PD and other neurological disorders. The door opened a bit, and we could hear the conversation between the lady doctor and a male voice, presumably the lead doctor. They seemed to be discussing my question, including something about referring me to another specialist. I was sorry I had said anything about the saliva and runny nose, a subject that eventually went unanswered. I just wanted to get on with the exam so that we could head back home before the rush hour traffic got too bad.

The door opened and the new PD specialist entered

the exam room. He gave me a quick handshake and he asked how his "old buddy" Dr. Rao was. After that, he was almost silent. The new doctor ignored Jeanne. *(In contrast, Dr. Rao always asked Jeanne, whom he addressed as "the Boss," how my symptoms REALLY were?)* The assistant did not join us. The specialist had me walk down the hallway and back, checked the flexibility of my arms and hands, and he made a note regarding my Requip dosage. Then he picked up the phone and dictated his observations into the phone, which seemed to connect him with a recording system in the Clinic. I overheard him say that I had a good gait when I walked, but I was stooped.

Without so much as a "be seeing you," he got up and left. "What do we do now?" I asked Jeanne. She shrugged her shoulders. I found a biographical pamphlet in one of those "take me" racks on the wall which summarized the doctor's training and experience. He specialized in Emergency Room work. Oddly, no mention was made of Parkinson's disease. The biography indicated that he had been associated with a medical facility in New Orleans. Perhaps that assignment was where he met Dr. Rao.

Finally, a nurse came in and gave me a prescription for more Requip and said the neurologist wants to see me again in five months. We were escorted to a patient checkout area, where the next appointment date and time were provided. And that was it, except for an unpleasant drive back to Jupiter.

Just what did we learn from the new specialist? I hated to say, not much. We liked his assistant more, since she at least communicated with us and seemed to be interested in my case. But the lead neurologist seemed to dismiss us as unimportant. I don't think he liked us. He was no Jay Rao. We missed Dr. Rao.

But we had a more immediate crisis—how were we going to get back home in the nightmare traffic? Jeanne called Chris and said, "If you can't find us a better way back home than 595 and I-95, then you will have to come and get us, because I absolutely will *not* go back the way we came!!!" Not wanting to do that, Chris said, "Don't worry, Mom. Stay by the phone." In a couple minutes, Jeanne's cell phone rang, and a Hertz employee in a nearby Hertz office gave her directions from Weston to the Sawgrass Parkway to Florida's Turnpike. This route proved to be much better and we took that route whenever we traveled to Weston.

I continued to take Requip three times a day. I went to the gym and worked out with my trainer every week. And I attempted to get back to something like normal in my daily activities.

But progress was slow and I was struggling with the symptoms of PD. If I sat down to rest, I fell into a deep sleep and ended up waking up as I drooled onto myself. I still had trouble walking. For example, I was very apprehensive that I might stumble and fall in the Communion line at church. I was stiff and uncoordinated and my balance was poor. My personal trainer constantly had me working on safety in walking and improving my sense of balance. But she took no chances; she held onto me with a fabric tether strapped around my waist to prevent me from falling. Since she is small, I hoped that I didn't fall or we both might hit the deck. But with regard to my getting hurt in one of her sessions, she said, "Not on my watch!" I felt secure under her supervision.

I had lost much of my manual dexterity and strength. This was evident in a restaurant, where I could not cut my own meat. I seldom finished my dinner when we ate out, because I was slow and also my appetite was poor. I also had trouble drinking from a glass, but we found that using a soda straw helped. Getting up from the table in a restaurant was a risky proposition. I had much trouble pushing my chair back, standing up and walking out of the restaurant through the maze of tables, waiters and patrons.

I had a good laugh when a group consisting of Jeanne, Betsy, Chris, Monica and I had dinner at a nice restaurant in Jupiter on Mother's Day. I was seated on the inside of the table, away from the aisle. Everyone was in a jovial mood as the group got up to leave. But they forgot something—ME! They were laughing and talking as they paraded out, leaving me trapped by the table and heavy wooden chairs. I fought hard to get free without falling. Diners at nearby tables were watching me with concern. Then I saw her. Monica had spotted me and was laughing her head off, as was I too.

“Michael, Michael,” she said in a playful but commanding tone. “We better watch you more in the future.” She rescued me and led me out of the restaurant where the rest of the group was waiting with big grins on their faces. Oh well, everyone loses something once in awhile. I was happy that Monica had noticed that I was missing. But I was embarrassed by the incident. Where was I headed, down the tubes?

The Author, Nov. 2004, Requip

All of this is to say that Requip was not a magic pill for me, but it offered the best I could hope for at that time. I continued to work hard and vowed to not surrender to this disease.

I visited Cleveland Clinic again, but we also kept in touch with Dr. Rao in New Orleans. Things sometimes go down before they go back up. Just when I thought my PD symptoms were improving, I realized that improvements were hard to come by, and I had begun to slip again.

Everyone has ups and downs in their life. That is to be expected. The photo above was taken on Thanksgiving Day, 2004. I was miserable and looked it too. I was drowsy, nauseous and exhausted. But in my case, and so many other similar cases, the "downs" are severe and hard to reverse. This was my problem with PD. I knew that Requip was supposed to be effective for PD, and I really was waiting for it to take hold. But that simple turnaround seemed elusive. I was still weak with an advanced case of PD. I did not have tremors, which was a huge blessing. But I was having trouble getting back my ability to do even simple things around the house. So I began to question the effectiveness of the medication and turned to the Internet for information.

GlaxoSmithKlein, the drug's manufacturer, had a detailed scientific discussion on their web site, with chemical formulas, and the results of a lot of human trials of the drug. They didn't know for sure how Requip works in PD cases. However, it was believed that Requip stimulates the dopamine receptors in the brain.

Another impressive benefit from Requip is its long term effectiveness. GSK presented data based on clinical studies that showed Requip effectiveness declining only about five percent over five years. On the other hand, the dopamine substitutes, carbidopa/levodopa, declined in effectiveness by 25-30 percent over a similar time span.

We trust Dr. Rao completely, and he believes in the value of Requip. He said that Requip protects the brain.

Nevertheless, I was beginning to have some doubts about my improvement prospects. It concerned me that so many chores around the house were being done by Jeanne—chores that only a few years ago I performed. When she was working in real estate I prepared some of the meals, I vacuumed and cleaned the house, I drove myself, I helped with real estate support tasks, I took care of the spa, etc. After I slowed down, she was under a lot of pressure to keep the household running.

We needed a fresh evaluation from our PD specialist in the Cleveland Clinic. He had us on a five month visit interval, which meant that we saw him effectively about twice a year. In 2005 we drove to Weston for an office visit twice, in February and September. The PD specialist was becoming increasingly distant. He would walk into the examination room and not even say hello. He did not look at me. He asked no questions. His only useful function seemed to be to write a new prescription.

HURRICANE KATRINA HITS NEW ORLEANS

We decided that we needed to find another PD specialist, but we were distracted by events in New Orleans. Stacey was pregnant, and her projected due date was in mid-July of 2005, right in the thick of one of the most active tropical storm and hurricane seasons ever recorded. Several violent hurricanes threatened the Gulf of Mexico area in July. Stacey's obstetrician was concerned about delivery of a baby into the teeth of one of these storms, so a C-section was performed on July 9, 2005, and Olivia Grace Stanfield was born, our first Grandchild.

Mother Nature had another nasty surprise for residents along the Gulf Coast as Hurricane Katrina formed over the Bahamas in late August, crossed Florida as a Category 1 Hurricane, strengthened quickly to a Category 5 Hurricane in the Gulf of Mexico and weakened somewhat to a Category 3 Hurricane before making landfall August 28th in southeast Louisiana. The levee system in New Orleans failed because the Federal Flood Control facilities were breached, causing 80 per cent of the City to be flooded. Mike and Stacey had bought a condo in the French Quarter so they were at risk as Hurricane Katrina moved inland. Tom Benson, the owner of the Saints, ordered players, coaches and office personnel to evacuate immediately before the roads became impassable.

Mike called us from his cell phone, and said "Ma—Stacey, her Mom, the baby and I are leaving to go east to Alabama. We also have the dog and Mom's bird with us."

Jeanne responded that we are terribly worried, are they safe? Mike said that they are in bumper to bumper traffic on I-10, but they plan to go north to skirt the storm and flooding. He will call later when they get to their condo in Gulf Shores (Alabama).

We did not hear from them for hours on end. Jeanne and I agonized and imagined the worst. Finally Mike called again and said that they had given up on Alabama as a refuge, because some roads were closed and the storm damage was too extensive. He had a favor to ask: would it be OK if he brought his family to stay with us in Florida? Of course, our answer was yes. About a day later the refugees from Katrina arrived at our house, tired but safe. Mike was called by the team and ordered to return to Louisiana to supervise the temporary relocation of the Saints team to San Antonio, Texas. Trish, Stacey's Mom, stayed in a motel in Jupiter with the bird and the dog (Jeanne is highly allergic to birds). The storm refugees stayed with us for about a month until Mike found a temporary residence for them in San Antonio.

What happened to our favorite doctor - Dr. Rao? Was he OK? Was his office intact? Was the LSU Healthcare Clinic still standing? Dr. Rao's office phone number was not working. How could we get in touch with him? We needed to see him. Jeanne and I both missed the know-how of Dr. Rao. Quite frankly, he was the only doctor we trusted. But for now, we would have to forego seeing him because of the storm damage.

In early 2006 we decided to find another neurologist in the Jupiter area, who could hopefully help me more than the doctor in Weston. At least Jeanne would not have the burden of driving to Weston. So we started over again with still another general neurologist in Jupiter. She seemed knowledgeable, pleasant and supportive, but nothing special regarding PD. She offered nothing new to help reverse my decline. She had no history on me and had no way of knowing whether I was rising or falling. But I was failing noticeably. I was miserable. I needed help.

Jeanne called once again upon our emissary in New Orleans - our son, Mike. "Mike," she said. "We just *have* to find Dr. Rao. Can you help us?" Mike replied "Ma—give me a couple of days, I'll get back to you."

Since this is a story with a happy ending, you can guess that Mike somehow tracked down Dr. Rao. Yea for Mike! The good doctor had moved his practice to Baton Rouge, Louisiana, but he was seeing patients one day a week on Fridays in New Orleans. He was in a different location, a medical facility on Prytania Street in New Orleans. We were able to make an appointment to see him in May, 2006. Jeanne and I were very pleased to reconnect with Dr. Rao. Chris went with us to assist me when we changed planes, and to drive us to Baton Rouge should that be necessary. I felt encouraged that we were doing something positive to stop my decline.

So off we went, via Continental Airlines to Houston, changing planes to fly into New Orleans at about noon on a Friday. I was tired, but still ambulatory. As I struggled to walk off the plane, I was met at the exit door as I departed the plane by a uniformed attendant with a wheelchair. He said that I should get into the chair. I said, "No thanks." He insisted, *"Sir, you should get in the chair."* I said NO. He said YES. Then Jeanne came back down the ramp and she also said YES! As I got into the chair I realized that I had sunk to a new low. My vitality was slipping away. When a complete stranger thinks you need help, you are pretty far gone.

Was I discouraged and demoralized? Yes. This wheelchair ride was embarrassing to say the least. Chris went on ahead to collect our bags in Baggage Claim. We planned to meet him outside on the ground floor level where we could take a shuttle bus to pick up a rental car. Jeanne gave the attendant a tip as he left us at the rendezvous point. My wheelchair rested on an inclined sidewalk that sloped downward toward the street, where cars circled to pickup arriving passengers. Jeanne held onto the wheelchair, but just then she spotted Chris coming with the bags and, to help him, she let go of the wheelchair, which of course started to roll.

“WHOA!” I exclaimed. “Jeanne - how do I stop this thing?” She did not hear me because she was talking to Chris. The chair picked up speed and then it hit a large support column next to the road. BAM! It came to an abrupt stop. With rubbery knees I made my escape. Thankfully I didn't get hurt. No one had seen me joyriding in the wheelchair. Good thing. I wanted to forget it anyway. How embarrassing.

The drive from the New Orleans airport to Prytania Street took about half an hour, and fortunately our Mapquest directions were accurate.

"Boy! I sure hated that ride in the wheelchair," I complained to nobody in particular. "I don't ever want to be in a wheelchair again!" No one heard me. "How did that wheelchair guy spot me?" I continued. "Was he tipped off by a Flight Attendant?" Jeanne finally said that I should just get over it and she ordered me to "*stop complaining.*" She was tired too and I was a handful. It was about 1:30 p.m. on a Friday in May, 2006. We had an appointment with Dr. Rao at 2 p.m. As I mentioned, he had relocated to a medical complex just outside the downtown area in New Orleans, where he saw patients one day a week. We found the medical building on Prytania Street, and Chris parked in a nearby garage. I was exhausted from the trip and still in an emotional tailspin as a result of the forced ride in the wheelchair.

Chris went off to find something to eat, and Jeanne went with him (probably to get a rest from me), while I checked in with the receptionist and took a seat in the waiting room. I tried to look normal, but I felt terrible. There were several other patients who also looked terrible. Had I joined the Land of the Living Dead?

Jeanne and Chris returned with a large bag of fried chicken and French fries, but there was no time to eat...

— 7 —
FINE TUNING

DR. RAO – NEW MEDS

A door opened and inside the hallway a friendly face appeared with a big smile—it was Dr. Rao, who motioned for us to follow him. I got up as best as I could, but I could not smile and walk at the same time. I was experiencing a collapse in my balance and coordination.

Dr. Rao's smile vanished and he looked shocked when he saw me approaching him. He turned to Jeanne and said, "What happened to him? Did he stop taking his medicine?"

Before Jeanne could reply, he turned to me and said, **"You look *TERRIBLE!* What happened to you?"** Jeanne, Chris and I followed Dr. Rao into an exam room. "You need new medication," he said. "I'll be right back."

He returned carrying a small packet of a prescription drug called Parcopa. He explained that this was a powerful form of levodopa-carbidopa. Be careful, he said, not to handle one of the tablets with wet fingers, because it would disintegrate. Four times a day I was to place one tablet on my tongue. It will melt away. He instructed me to let the melted tablet remain on my tongue for a minute, and then swallow it. "Do it now," he said, handing me a small tablet in a sealed packet.

I did exactly what Dr. Rao said. It was the most interesting medicine I ever took. It dissolved immediately. Wait—wait—wait - then swallow. Well, what next?

Jeanne, Chris and Dr. Rao watched me intently. "How do you feel?" he asked. I felt like a goldfish in a tank, or an animal in a zoo. But he meant my response to Parcopa.

I replied, "It's incredible! I feel better already." The drug must have gone straight to my brain. Astonishing! WOW! I'm going to recover.

Dr. Rao explained that this "shock treatment" was just to get me started on a different program of medication. He gave me two prescriptions, one for just a few days of the "power pill", and another for long term use, called **Stalevo 100**. I was to continue taking Requip as before.

> ***Stalevo 100*** contains carbidopa, entacapone, and levodopa, which are medications used to treat Parkinson's disease. Parkinson's disease is believed to be related to low levels of a chemical called dopamine in the brain. Levodopa is turned into dopamine in the body. Carbidopa and entacapone are used with levodopa to prevent the breakdown (metabolism) of levodopa in the body.

We had a pleasant chat, during which Dr. Rao mentioned that his daughter, also a doctor, was about to deliver his first grandchild. Jeanne told Dr. Rao that we would be back to see him every three months, the next time in July, 2006 when Olivia would be one year old, and also would be baptized. He was delighted. We were delighted. Dr. Rao asked whether we would be staying with Mike and Stacey - we said yes, and he said he would call us on Sunday to see how I was doing. *(And true to his word, he did call us on Sunday.)*

Away we went in the rental car, on our way to see Mike, and Stacey and the main event: Olivia. I felt better after seeing Dr. Rao. I even ate some cold French Fries that Chris didn't eat. But first we needed to get my prescriptions filled. Was that easy? NO! We went to FOUR Walgreens Drug Stores. None of these stores had these drugs - of course they could special order them, but they would not be delivered until Monday afternoon, when we would be on our way back to Florida. Finally we found a CVS Pharmacy that could fill my prescriptions. Chris was great, negotiating the unfamiliar highways and finding so many drug stores. *(Lesson: PD meds can be hard to find.)*

We had an enjoyable weekend at Mike and Stacey's new home in Kenner. They had sold their condo in the French Quarter in New Orleans. It had received only minor damage from Katrina, but they wanted a suburban location for raising a family. Olivia was lots of fun. One evening we all had dinner at Morton's Steak House. Mike Jr., Stacey and Chris paid a late night visit to a Casino in New Orleans, which caused Super Mom (Jeanne) to sound Battle Stations when they did not return by 2 a.m. They showed up just as we were about to start calling the police and hospitals.

I was already getting better, much better. Nobody offered to give me a wheelchair ride on our way back home. In our May visit to New Orleans, we told Dr. Rao that we planned to see him about every three months, the next time in July, coinciding with Olivia's first birthday and baptism.

And indeed, Jeanne, Chris, Monica, and I traveled to New Orleans the second weekend in July, 2006. We stayed in a hotel near the airport, a convenient location not far from Mike and Stacey's new home in Kenner. Their guest bedroom was occupied by Joe and Carroll, Olivia's Godparents.

In the minor glitch department, during our July visit we got lost going from the airport to Dr. Rao's office on Prytania Street. We had a printout from Mapquest, but somehow we made a wrong turn and got hopelessly lost in poor New Orleans neighborhoods that had major damage from Hurricane Katrina. We tried to figure out where we were on a street map of New Orleans, but did you ever try to read an unfolded map with microscopic printing in a moving car on narrow streets while simultaneously looking for street signs? Some of the street signs were missing, probably blown away in the hurricane. Around and around we went. Boy! Were we lost! I was concerned that I might miss my appointment with Dr. Rao, so Jeanne called his office and said we would be a bit late. Finally Jeanne called Stacey who was in a store at the time. Someone in the store overheard her conversation and offered some directions, which were very helpful. All of a sudden, there it was—a street we recognized and we found the medical center at last.

Chris went for something to eat, and Jeanne and I

entered Dr. Rao's reception room. It seemed to be deserted . . . oh-oh, did we miss him? But no, he was still there and once again his pleasant voice could be heard beyond the waiting room. The door opened, and out he came, all smiles, especially when he saw how much I had improved since our visit in May 2006. He advised that I should continue the same treatment. We told him that we planned to visit again in October 2006, and we would call for an appointment. He said that we should have Mike make the arrangements.

It was a great weekend, with a huge first birthday party for Olivia on Saturday, who loved her presents, and who celebrated by covering her face with icing from her birthday cake.

On Sunday morning we all went to Mass, attended by Olivia, her parents, Godparents, two grandmothers, one grandfather (me) and Monica and Chris, plus assorted friends. After Mass, Olivia was baptized in the church. She stole the show, looking radiant in her baptismal dress. We had dinner at Impastato's (restaurant), normally closed on Sunday, but open that day just for us in honor of Olivia. We were serenaded by a vocalist while we ate dinner. All in all, it was a dream weekend. And I was continuing to function much better under the new medication regimen prescribed by Dr. Rao. I joked that I was almost a Real Person again.

Week by week we noticed a gradual improvement in my symptoms. I was getting stronger, better coordinated, with improved balance and flexibility. The Stalevo 100 was truly a wonder drug for me. It seemed to increase in effectiveness with continued use. One measure that I found useful was how I performed on an exercise bike in the gym. Before Stalevo I could pedal for only about five minutes at 35-40 RPM before becoming exhausted. After Stalevo I could pedal at 50-70 RPM for twenty minutes and beyond. I had a renewed interest in going to the gym - exercise was beginning to really pay off.

It was like getting your life back after being told your condition is fatal. Boy! Was I Lucky! So, after returning home in July 2006, we started to develop plans for a family reunion weekend in New Orleans in October. Chris and Monica planned to go with us. We were doubly delighted that Betsy and Keith would also join us for the weekend in New Orleans. To avoid getting lost in New Orleans again, I added still another travel companion, **Nüvi**, a Garmin GPS unit. A Family Reunion! Yea! The weekend starting Friday, October 13, 2006 was chosen to coincide with a Saints home game.

The Author - August 2006
Stalevo 100 and Requip
FEELING GOOD

Just about everything that had been adversely affected by PD was improving. Some improvements happened so smoothly that it was days or weeks before we noticed that I was better. For example, before Stalevo I had so much trouble drinking from a glass that I used a soda straw, but after Stalevo I found that I did not need to use a straw. I stopped using a straw after I realized the glass was easier to handle *without* the straw.

We also needed an appointment to see Dr. Rao, so Jeanne called the LSU Healthcare Clinic at the Prytania Street address and requested an appointment on Friday, October 13, 2006.

The receptionist, who sounded disinterested, asked “Which Doctor?”

Jeanne replied, “Dr. Rao.” *Dead silence.* Jeanne repeated her request . . . *more silence.*

"Dr. Rao does not work here anymore," was the eventual reply in a dismissive tone of voice. *WHAT?* Jeanne took a moment to digest the meaning of this statement. "Where is he now?" she asked.

The receptionist shot back, "He retired." Stunned, Jeanne hung up. Well, this news came as a complete surprise. Where had he gone in retirement? — *that is,* if he really did retire. I went back to the Internet to search for Dr. Rao. No such luck. I found the same old postings, and a bunch of phone numbers which I recognized as being out of date. We did not have his cell phone or home phone numbers. Who was best able to find Dr. Rao? Of course, our son, Mike, who, as expected, tracked him down. Mike found that Dr. Rao had relocated his practice to Ochsner Clinic, which is a large medical facility in the outskirts of New Orleans. The travel distance from Mike’s house to Ochsner is less than the distance to the LSU facility in downtown New Orleans, which was a plus. Olivia had been born at Ochsner, so Stacey and Mike were already familiar with the facilities. Mike made the arrangements for our visit to Ochsner, and gave us the information we needed. Great news! We called to confirm our appointment and breathed a sigh of relief - Dr. Rao was scheduled to see me at 11 a.m. on Friday, October 13, 2006. (*Lesson: Keep in touch with your Doctor.)*

We were pleased to learn that Southwest Airlines had begun a non-stop service between Fort Lauderdale and New Orleans. This direct route reduced air travel time by an hour, eliminated the hassle from changing planes, and reduced the risk of lost luggage. We also arrived with less travel fatigue. This route did require a fairly long (one hour) limo trip between our home in Jupiter and Fort Lauderdale. Also, the airport in Fort Lauderdale is much bigger and busier than the airport in West Palm Beach. Nevertheless, we like the non-stop service best.

STILL MORE MEDS – A PATCH NO LESS

In 2007 we traveled to New Orleans to be examined by Dr. Rao three times: in April, (when we watched over Olivia while her parents were on vacation), July (to celebrate Olivia's second birthday) and October.

The birthday party in July 2007 represented a significant turning point in my treatment for PD. Of course, as usual, Olivia and her pint-sized guests had a wonderful time with toys, birthday cake, ice cream and paper hats.

In the early evening, as the party began to wind down,

the adult party goers split into two groups, inside the house, and outside, on the back patio. I was an insider, getting a bit tired, just relaxing in a chair in the dining room. Suddenly Mike Jr. rushed into the dining room and said, "Dad - Dad! Come quick, they are talking about Parkinson's." I got up and hurried after him to the patio, where a group of about six people were seated, chatting about something to do with Parkinson's disease. They were listening to a guest who was telling the group about a certain Dr. Rao, a world-renowned PD specialist, who was assisting a pharmaceutical company in testing a new drug for the disease. He said it was a new 24-hour patch. I mentioned that Dr. Rao was my PD specialist too, and I was encouraged that he is so famous. He really must be top notch. I also was intrigued by the patch.

It was beginning to get dark. A guest asked Mike if he had any Diet Coke in 2-liter plastic bottles. The guest explained that there is a fascinating experiment — if you drop Mentos brand peppermint candy into an open, full bottle of Diet Coke (not Pepsi), the fizz in the coke is released so fast that the Coke is vigorously expelled straight up, sometimes as high as 15 to 25 feet. Mike checked and found that the Diet Coke supply was about out, so he paid a visit to a 7-Eleven and stocked up.

The crowd hushed as a full 2-liter bottle was positioned - cap off - about 20 feet from the house. Mike slowly approached the bottle and dropped a Mentos candy into the bottle. The crowd sighed in disappointment when the coke barely burbled up — a flop. Much discussion ensued as powerful minds worked on the problem. Again and again they tried, and found that increasing the number of Mentos produced a better Coke launch. The crowd cheered the brave volunteers, the "Mentos Droppers," who courageously got soaked with Diet Coke as it fell back to earth. GOOD JOB! The great birthday weekend was over. And we had a few laughs too.

Our July 2007 visit to New Orleans also included an appointment with Dr. Rao a day after Olivia's birthday party. This exam provided an opportunity to ask him about the patch. He described the patch as being able to deliver to the patient a uniform amount of medication over a 24-hr period, as we learned the day before. It would replace the Requip that I then used. He said that we should make an appointment to see him around February 2008. However, he was not sure when it would be available, but approval by the FDA had been secured by the manufacturer. Dr. Rao was pleased with my progress and recommended that I continue the treatment and exercise as before.

The summer months in 2006 and 2007 were quiet on the weather front, which was a welcome change from the wild hurricane year in 2005. We had an appointment with Dr. Rao on October 30, 2007. He thought that my condition had slipped a bit, but recommended that I continue treatment and exercise as prescribed. We discussed the timing and amount of Stalevo 100, because I explained that I had noticed some slippage and I was having some discomfort from side effects. Dr. Rao said that I could take Stalevo every four hours, six times a day.

We also discussed several other symptoms, including poor posture, and some unusual vision problems. On our recent flight to New Orleans the row of seats just ahead of us appeared to me to be sloped down from right to left by about four inches. Jeanne said no, the seats did not slope down. Dr. Rao smiled and said that this aberration is a well-known phenomenon in the brain of PD patients. Sometimes I have trouble placing an object, like a water glass, in the desired spot on a table. I also have trouble tossing an object, like a crumpled piece of paper, into a waste basket. I almost always miss the waste basket. I have still another unusual visual oddity - sometimes, when I move my head, objects in my field of view stay still and then catch up with the background.

I told Dr. Rao that I sometimes cough while asleep and wake myself up. I thought that this cough was due to excess saliva getting into my breathing passages. Dr. Rao suggested that I also may have a dry mouth condition, which can cause coughing at night. The treatment— chew gum. This should help to keep the mouth moist and stimulate swallowing. My trainer is having me work on swallowing exercises.

We also discussed the swelling of my ankles. Dr. Rao said that Requip can cause edema in extremities. He said that I should elevate my feet, which should help.

I saved the big news for last. Dr. Rao said that he had good news. The patch was available! He explained that it is called **Neupro**, and it comes in 2, 4 and 6 mg strengths. One patch is applied per 24 hour period, during which it delivers the indicated amount continuously. He gave me some samples to try, in the form of a dosage kit for determining the proper dose. He said that I should not bother trying the 2 mg strength. I will probably find 4 mg or 6 mg to be best. He gave me a prescription for the 6 mg patch, which he thought would be best for me.

I waited to get home before starting my first application of Neupro. When I did start, I was really surprised that this medication and delivery method was so very, very good! Within two hours, I FELT BETTER. MUCH BETTER! Wow, this medication was essentially free of side effects. No more nausea. No more drowsiness. No more depression.

I decided that 4 mg was satisfactory, and Dr. Rao's office faxed a prescription for that amount to the pharmacy, which obtained the 4 mg patch and rushed it to me overnight.

Neupro is a new skin patch that contains the drug rotigotine. It may also be combined with levodopa in patients with more advanced cases of the disease. This medicine helps dopamine receptors in the brain work better. The patch is replaced every day. The medicine is delivered through the skin uniformly over a 24 hr. period.

This treatment should improve my quality of life. It should open up new horizons. Can I drive more? How about stamina? Will I be able to work, play and exercise more without getting tired? This exciting development is evidence that PD patients should not become discouraged. I feel certain that even more improved treatments will be coming in due time. I agree with those who say that the drug companies and universities should accelerate their research to find a cure for Parkinson's.

Here is an example of serious work underway to solve the puzzle of Parkinson's disease. In an article in the May-June 2008 American Scientist magazine *(The Neglected Side of Parkinson's Disease),* neuroscientists *Rothstein and Olanow* suggest that Parkinson's disease may be the product of other brain abnormalities that interact to kill brain cells and thus cause the classic symptoms of PD. They postulate that it may be possible to identify and treat these disorders early enough to prevent Parkinson's from getting started. I hope so!

2007 -- A VERY GOOD YEAR INDEED

Without doubt, 2007 was a very good year for me. My quality of life did improve significantly with changes in medication and continued exercise. We went to New Orleans for Christmas 2007, for the first time *without* the need to see Dr. Rao. I was making progress.

The most recent visit to see Dr. Rao was in February 2008. He answered several questions, saying that a minor chattering of my lower jaw, which was so small that it was nearly impossible to see, was nothing to be concerned about. He also said the edema in the area of my ankles was due to the medication. He recommended elevating my feet and wearing different socks to minimize the swelling. He was in a good mood, pleased with my progress, and delighted that Jeanne and I had our cute-as-a-button granddaughter and her mom, Stacey, with us. Olivia, as usual, stole the show. Wasn't I lucky to have improved so much that we had time to enjoy our "grand babies?" Dr. Rao showed us his iPhone loaded with photos of his granddaughter.

"See you in about five months," he said. Jeanne responded that we would be back in June (a four month interval), when Olivia would be in a Dance Recital. "OK. Fine, See you in June then," he replied.

Dr. Rao is so much in demand that he's like a Rock Star, booked for months in advance. As soon as we got home we made an appointment for July 2008, the earliest we could get. ***Amazing***!

THE CUTTING EDGE

Being on the "Cutting Edge" is great if one wants to have the latest technology, in this case, the Neupro patch. I am delighted to have had an opportunity to use it. In my experience, this patch delivered most of the medication so uniformly that I was unaware that I was using a patch. Apart from some annoying skin irritation, the Neupro delivery system seemed to perform well at first.

Well, you can imagine my disappointment when I received a letter in early 2008 from my mail-order drug company, MEDCO, saying that the manufacturer of Neupro had temporarily discontinued this medication due to quality problems. The letter went on to say that all supplies of Neupro were expected to be exhausted by the end of April, 2008. Therefore, patients using Neupro were advised to contact their doctor as soon as possible to withdraw from Neupro and substitute a different medication.

So I called Dr. Rao's office and learned that possibly only a few batches were defective. Nevertheless I was advised by Dr. Rao to substitute Requip tablets until Neupro production resumed (if ever).

But better news was on the way—*Requip in a time-release form.* At first, I was apprehensive about resuming the use of Requip, which caused unpleasant nausea, drowsiness and depression. But the time-release feature might dampen these side effects, so I eagerly awaited the debut of Requip-XL.

Sometimes "Being on the Cutting Edge" cuts both ways. If this is the price of new developments in treating Parkinson's disease, then COUNT ME IN! However, one should expect occasional setbacks.

WHERE AM I HEADED?

Well, that brings my story up to date. I have a good degree of control over my life now. The new patch and medications have helped me get a grip on my physical activities. I have a positive outlook. My exercise program is paying off.

Of course, nothing is perfect. In the back of my mind, I know that there is another level or two of decline possible in my Parkinson's disease symptoms. Decline into what? Well, I will deal with that if and when I need to deal with it.

So for now I prefer to cruise along with a positive attitude and work at holding my own. New medications are coming. I expect great things and pleasant surprises from Parkinson's disease research.

But what about the future? Where am I headed?

Well, I have thought long and hard about that question. I would love to tell you what I think about my prospects for

. . . THE BEST FUTURE POSSIBLE

In my dream of all dreams, the future will bring forth new developments in medical miracles that will cure Parkinson's disease.

That may happen in a year, or a decade, but I fully

expect that a cure will be found eventually. My guess is that the cause(s) of PD will be identified and preventative measures will spare future potential victims of this disease.

However, in the meantime, we are left to make the best of what we do have, which is nevertheless substantial.

NOTE FROM THE AUTHOR

By now you have learned that being a PD patient is hard work. Every day you must cope with difficulties brought on by the disease—everyday activities that most healthy folks take for granted. Even on the best days, you know that life for PD patients is extra tough.

Does that mean you have to bite your lip and suffer in silence? Not at all. There is much that can be done to improve your outlook. Part Three that follows is intended to inspire you to seek a better life. Good luck and work hard. It's your life. You are not alone.

They can because they think they can.

Virgil

Part Three

Tips for Living with PD

If you can DREAM it, you can DO it.

Walt Disney

— 8 — DREAM MAKER

I CAN DREAM, CAN'T I?

Throughout this book I have referred to my hopes and ambitions for the future. You may have noticed that whenever my conditions change, I try to assess how the change will impact my abilities to do what? - fulfill my dreams. Can I climb a ladder? Is it OK to drive now? Are my hands good enough to change that watch battery that sat on my dresser for two years? Will I be able to carry on a meaningful conversation with a store clerk?

You see, I have a "Tiger by the Tail." You must have a passion for something burning inside you too. Your dreams are not necessarily the same as mine. I wish to be a musician, a composer and an arranger of orchestral music, plus many more dreams. What's that, you don't have any dreams, because you have Parkinson's disease with tremors. Do you know what some amputees do? They run and finish marathons. *They have dreams.* And then there are artists without arms who paint beautiful works of art by holding a brush in their teeth or toes. *They have dreams.* There is a man who is paralyzed due to an accident, but he writes mystery novels. *He has dreams.*

But some PD patients seem to be in God's little waiting room, just waiting to die. The sooner the better, they say, because they think they would be better off dead. They complain and moan and sit around all day with long faces. And they are good at being sick. Sure, and they try to burden those around them with remarks like "Oh, I'm sorry. I don't feel well today. Where's the TV schedule?" The answer should be "go get it yourself."

Oh, you think that I must be exaggerating some. Not one bit. Some people can really be sad sacks. If you ask them to tell you a bit about their dreams, they rear back and say, "What good are dreams? I'm too sick to waste my time on dreams."

Well just one moment, PLEASE! We often see a U.S. soldier on TV who has been gravely wounded in battle and lost a limb. What is his dream? It is to return to his unit! Of course, the Army won't allow him to go back to his unit, right? Wrong. He is fitted with an artificial limb, goes through rehab and back to his unit. What does the soldier have that some PD victims need? COURAGE.

Webster's Dictionary defines 'courage' as a quality of

mind which enables one to meet danger and difficulty with firmness and valor. Courage stresses firmness of mind or purpose and the casting aside of fear. Now we have singled out a key factor in an incurable disease like Parkinson's disease: fear. Some patients are so afraid that they lose the ability to fight for themselves. The stress that this negative attitude generates then leads to a new form of illness. Studies suggest that the state of mind of people who are under stress damages the immune system. This damage is not just imagined, it is real physical damage. "Bad" body chemistry makes them tired, withdrawn, and exhausted. The efficiency of PD treatment declines and the patient's body engine goes into reverse.

Have you not heard of cases of serious illness, like cancer, in which the patient has an unexplained improvement? The local newspaper gets wind of a human interest story and sends out a reporter to interview the patient and the family. The patient, a middle aged woman named Jane, is eager to tell her story. She exudes positive ions. She can't wait to tell the reporter about herself. What story does she tell? About her battle with cancer? The drugs and medicines? The chemotherapy?

No, she tells the reporter all about her dream, which is to start a shelter for abused animals. She has a young dog named Bingo, who is lying at her feet, a beautiful black and white mixed breed animal. The dog seems so quiet, subdued, almost fearful. "This is Bingo, my sweet little dog," says Jane, patting the dog as if to reassure her that she is safe. The reporter leans down to pat Bingo on the head, but the dog pulls away.

"Oh, it's OK Bingo," says Jane, "This nice lady is not going to hurt you. You see, Miss Reporter, I rescued Bingo from our local pound, where they were going to destroy her. I think that this dog was abused as a puppy. She has physical and emotional scars. I don't know what was done to her. But someone really hurt her. Now I'm the only human that Bingo trusts. She is the main reason that I am interested in starting an Abused Animal Shelter."

So the reporter gets her story, but the cancer remission is only part of the story. The headline for the article which makes the front page of the Human Interest section of the Sunday newspaper reads "Local Woman, Recovering from Cancer, to Start Abused Animal Shelter."

And how is Jane described? As sick? No. As a Victim? No. Fearful? No. She is described as a person with a Dream. A Passion. A person with a fire in her belly. A Do-er. Courageous. She obviously loves animals. There is a physical response in the body to the positive feedback that Jane experiences as she makes plans for the shelter. Her love for her dog generates a glow in her that is obvious to the reporter. Difficult to quantify or even explain—this internal glow is a real effect that is described by some researchers as due to the body synthesizing naturally occurring chemicals called "endorphins." These substances are thought by researchers to reduce pain and sustain feelings of love, whether for a dog, or for one's dreams. They also seem to affect sensations of pleasure and reward, and could play some role in vigorous exercise and the so-called "runner's high" that helps a marathon runner to keep going under stress.

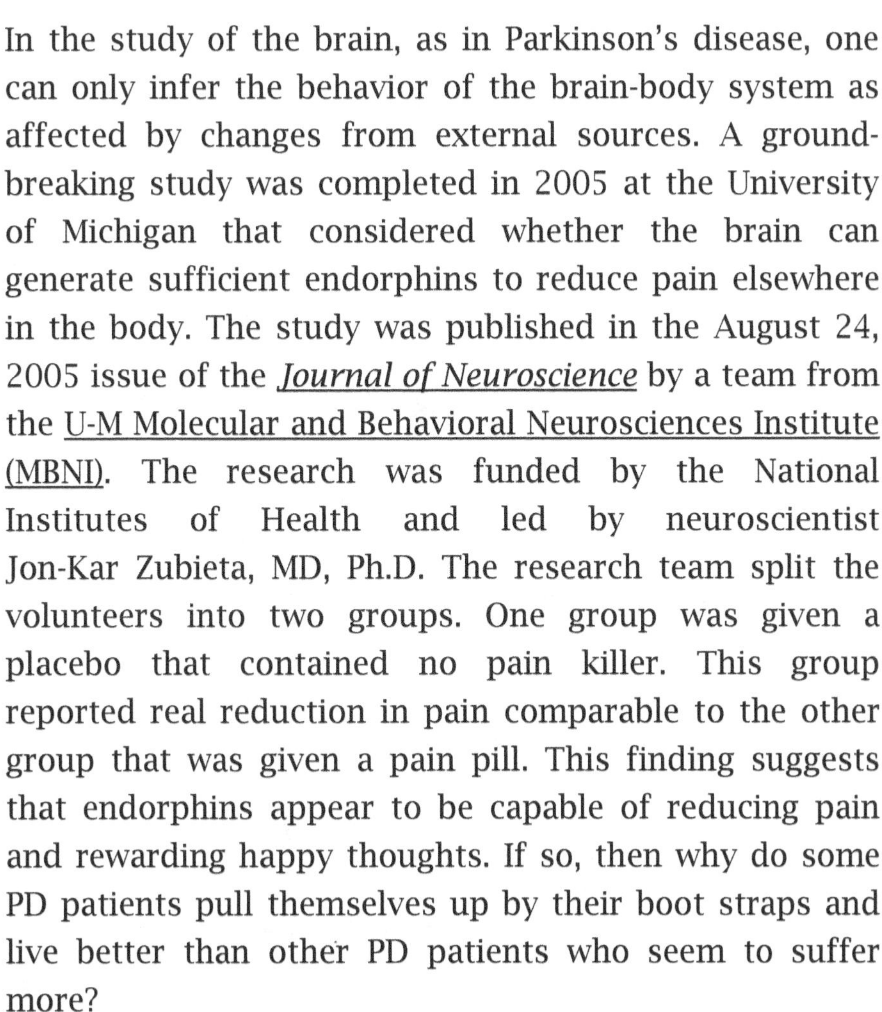

In the study of the brain, as in Parkinson's disease, one can only infer the behavior of the brain-body system as affected by changes from external sources. A ground-breaking study was completed in 2005 at the University of Michigan that considered whether the brain can generate sufficient endorphins to reduce pain elsewhere in the body. The study was published in the August 24, 2005 issue of the *Journal of Neuroscience* by a team from the U-M Molecular and Behavioral Neurosciences Institute (MBNI). The research was funded by the National Institutes of Health and led by neuroscientist Jon-Kar Zubieta, MD, Ph.D. The research team split the volunteers into two groups. One group was given a placebo that contained no pain killer. This group reported real reduction in pain comparable to the other group that was given a pain pill. This finding suggests that endorphins appear to be capable of reducing pain and rewarding happy thoughts. If so, then why do some PD patients pull themselves up by their boot straps and live better than other PD patients who seem to suffer more?

The answer may lie in differences in the disease itself as it manifests itself in different people, the patient's personality, life style, mental attitude or family support. In my case, as you may remember, my initial response to being diagnosed with PD was to retreat within myself. But shortly thereafter, my wife lit a fire under me. Mike Jr. went to work and found Dr. Rao. Within just a couple of days we traveled to New Orleans so that I could be examined by a leading PD specialist. In a few days I was on a strict program of medication. While one local neurologist said I should take an herbal remedy and come back in a month, Dr. Rao said I should start treatment *immediately.* When I expressed anxiety, Dr. Rao said, "Don't worry." When my symptoms got worse, new meds were prescribed. In short, I had a great deal of support and I was inspired to stop the progression of the disease if I could.

My exercise program was intensified: at first, two days or so a week, then four, finally six days (three with the trainer and three solo in the gym).

I declared - I WILL beat this thing, referring to PD as though it is a fire-breathing dragon that I must slay. I dare not look back. But don't be afraid. Don't get tired. Don't get discouraged. Don't slack off. If you do slip, pick yourself up and keep going. Go-go-go.

THINK ABOUT IT

Think. What a powerful concept—every human is born with a brain with which to THINK. Our brains perform many of the functions of a modern computer. There are inputs and outputs to and from our brains. Our brain processes information and applies complex rules to judge situations and choose courses of action. But we are actually *MUCH* smarter than a PC.

For example, just consider the driver of a car in fast rush hour traffic. In addition to getting to his or her destination safely, the driver is listening to MP3 music, talking on a hands-free Bluetooth cell phone, drinking Starbucks coffee from a travel mug and preparing for another day at work — all at the same time!

There are no computers that I know of that can do all of that on the spur of the moment. And, in the above example, with no advance warning the driver can pull the vehicle off the highway in case of a flat tire.

Wow! Where can I get one of those human computers? Hold on, my friend, you have one — the brain that you were born with. Use it and you may be surprised at what you can do to help in dealing with PD. Over the years researchers have sought ways to duplicate "human intelligence" in programming powerful computers. To the best of my knowledge, these attempts at creating artificial intelligence have not proved successful. But your intelligence is real, not artificial. And it's online now.

So it's up to you—use it or lose it—your brain, that is. PD may have already damaged your brain. Is it not time to strike back? Launch a counteroffensive. *Defend yourself.*

As a first step, consider carefully whether you agree with the treatment you are receiving. Don't just accept the recommendations at face value. In my case, the first three neurologists prescribed three different treatments. Hey, come on, they can't all be right. If the advice doesn't seem right, it probably *isn't* right. Learn and THINK about it. Take charge. Use your brain. Seek second and third opinions, if need be. Also, *never-ever* seek treatment from a family doctor not trained in PD.

Do you have a meaningful exercise program? If not, why not? Check out Appendix I (page 132), which gives examples of exercises for PD that you can perform at home — no expensive gym needed. Can't find a good personal trainer for PD? You can get a free exercise poster from the manufacturer of Stalevo. Contact Novartis for more information at 1-888-973-2666. You also can receive free tools to help manage your condition from Novartis.

Just THINK about it! You have the most complex and best personal computer in the Universe—the Human Brain. What did it cost you? Nothing, it's free for everyone, rich and poor alike. So use it! Learn new things. Apply your brain to accomplishing your dreams.

Think. Think. Think.

— 9 —
ACTIONS SPEAK LOUDER

STEPS YOU CAN TAKE

Here is a summary of my life with PD and how my actions so far have helped to reduce the effects of my symptoms:

1. Immediately after the initial diagnosis, seek medical help from the best Parkinson's specialist you can locate. Put everything else in your life on hold. Don't delay. Speed is essential. Your brain possibly is being damaged every day.

2. Do not rely on a family physician for treatment. However well-intentioned a general practitioner may be, a qualified neurologist is better suited by training and experience to help you choose an appropriate treatment. *See Appendix V (page 144) for help in finding a PD specialist.*

3. Take your medication exactly as prescribed. Deviations from orders can be harmful.

4. Don't feel sorry for yourself. Self-pity will only

slow down the recovery process. Instead, take heart, be brave and courageous. In your heart of hearts, know that you will survive this disease.

5. Start a vigorous exercise program. Get help from a qualified personal trainer, who will customize a program to fit your needs, supervise its execution, and adapt to changes in your condition. If you cannot afford a trainer, you may be able to have a physical therapist whose fees are covered in part by insurance. *See Appendix IV (page 142) for a resumé of Ms. Maribel Bleeker, a Professional Personal Trainer in the south Florida area.*

6. Be vigilant about changes in your condition, and report these changes to your PD specialist.

7. To the extent possible, try to be self-sufficient in your every day activities, including dressing, eating, and other personal chores. If you stop fending for yourself, you may gradually get worse until you need help constantly.

8. Discuss with your PD specialist the side effects associated with taking Parkinson's medication, such as drowsiness, that could contribute to an accident or injury. Do not drive a car or operate dangerous machines until your response to medication is determined.

9. PD patients sometimes have poor balance, which

can contribute to a serious fall and injury. Take special care to avoid falling.

10. Keep busy. Be active. Pursue a dream. Help your family to understand that you can deal with PD. Learn to smile a lot. Look for humor in everyday situations. A good laugh is good medicine.

11. Join a support group if this approach to sharing information appeals to you.

12. In order to document your progress, keep a written record of doctor visits, personal observations, weight, height and heart rate in exercise.

Above all, keep your spirits up. Remember, you are not alone. Have courage. Get help. Talk to family and friends.

DISCOVERING THE REAL YOU

Just who are you? No one really knows who they are until some life-sized event or change in their life path occurs. And what is a "life path?" The concept of a life path may be new to you, but we all have one, it's just hard to see sometimes, especially when we are immersed in day to day problems.

An example of a change in one's "life path" occurs on a wedding day, when the bride and groom exchange marital vows. The wedding couple is still composed of the same minds and bodies as before the wedding, but a new facet of their "life paths" is revealed to them first of all, and to all who know them later. The change in "life path" can be for the better or worse, in sickness or in health, so the vows go. If the groom is fundamentally a "good guy," his life path leads him in a new direction that reveals his capacity to love and sacrifice for his wife and family. Why could we not see this capacity before they are married? Because the pressures of married life are different than before the marriage, so a change in life path brings forth the new dimension of the married couple for all to see.

Another example of a "life path" is found in the response to learning that an individual has a serious illness like PD. Now I can tell you from personal experience - the initial diagnosis of Parkinson's disease was truly a shocker. My trolley car came off its track for a while. I truly had to get a grip on my emotions. But more than forty years of honing my "life path" paid off, and I was able to get back on track. This initial success resulted in my steering down the path I intended and I gained confidence that I would be a PD survivor, not a PD victim. I became more enthusiastic as a promoter of treatments and eventually a cure for the disease.

We need to discover our true inner self, a vision of the real me and the real you. Nothing will accomplish that unmasking any more positively than living with and moving on with Parkinson's disease. Accept the challenge and move on. You will find the real you. You will be proud of yourself. And your family and friends will be proud of you too.

DREAMS OF THE RICH AND POOR

Discover your fondest dreams, those real dreams that lie ahead, whether you are young or old. Follow in the footsteps of others of all ages. History is loaded with examples of people who have achieved remarkable accomplishments in spite of challenges.

One of my favorite comedians is George Burns, who began a second acting career at the age of 79. He continued acting until his death shortly before his 100th birthday. His challenge: longevity and advancing years.

John Wayne acted in 200 films, mostly westerns, in his fifty year-long career. His last film, *The Shootist* was about a gun fighter with cancer. Ironically, Wayne battled lung cancer himself. His challenge: declining health.

Small in physical stature, but a giant in faith, Mother Teresa of Calcutta tended to the sick, hungry and impoverished poor for many of her 87 years on this Earth. She was awarded the Nobel Peace Prize and shortly after her death in 1997, she was Beatified by the Catholic Church, a big step toward Sainthood. Her challenge: poverty and advancing years.

Thomas Alva Edison was born in 1847 and was awarded a remarkable 1,093 U.S. patents before his death at the age of 84 in 1931. His laboratory is on display in Menlo Park, New Jersey, where he invented the carbon filament electric light bulb, telephone transmitter, phonograph record, motion picture camera and electric power distribution system among many other devices that are essential today. His challenge: competitors.

Michelangelo was an incredibly talented painter, sculptor, architect, poet, and engineer who was born in 1475, and devoted most of his 89 years to masterpieces such as the statue of David, the Pieta, the Last Judgment (painted on the ceiling of the Sistine Chapel) and the design for the dome of St. Peters. His challenge: political favoritism and betrayal.

Dubbed America's all-time finest Architect, Frank Lloyd Wright designed over 1,000 projects, of which more than half were built, the last one his masterpiece, the Guggenheim Museum in New York City. He was 91 years old when he died in 1959. His challenge: advancing years.

Here are more examples of famous elderly whose bravery and hard work led to fulfillment of their dreams: Coco Chanel, Cecil B. DeMille, Gandhi, Justice Holmes, Archibald MacLeish, Margaret Mead, Golda Meir, Claude Monet, Pablo Picasso, Helena Rubinstein, Artur Rubinstein, Bertrand Russell, Albert Schweitzer, George Bernard Shaw, Casey Stengel, Leo Tolstoi, Giuseppe Verdi and Mae West.

A special recognition should go to Pope John Paul II, whose courage and fortitude when faced with Parkinson's disease surely is a model for us all, with or without the disease. This is a good time to discuss the next topic, using the Pope's illness as an example of courage:

THE SPIRITUAL DIMENSION

When it is a matter of life and death, many people look for help to their spiritual side, which I call the "Spiritual Dimension." If you are suffering with PD, take a close look at some of the video of Pope John Paul II addressing the faithful gathered in St. Peter's Square. Toward the end of his life, the Pope looked very weak in his "Pope-mobile," barely able to hold up his head, unable to speak clearly. The ravages of PD are painfully evident, although the Vatican did not confirm Parkinson's disease until 2003. In early 2005 the Pope fell seriously ill with influenza and was treated in a hospital in Rome. In late March he became gravely ill and passed away in his private residence on April 2, 2005. He never surrendered to PD.

As a source of inspiration and a reason to pray for help, I can recommend a CBS TV movie starring Jon Voight as Pope John Paul II. This film presents a compelling portrait of his life starting in Nazi-occupied Poland and ending with his illness and death in Rome. This robust and energetic man loved to ski and was a fine athlete as a young man. The movie provides a glimpse of the onset and progression of Parkinson's disease as it gradually took its toll on the Pope. This movie is available on DVD from online stores (Amazon, J&R, Blockbuster), under the title "Pope John Paul II" (starring Jon Voight).

What if you are not Catholic, or Christian, or Protestant, or Evangelical, or Jewish, or Muslim, or just any religion? Perhaps you don't even believe in God. Does my word no longer apply? Sorry, you can't get rid of me that easily.

Let me suggest that everyone needs to pray to God for help with Parkinson's disease. It is such a cruel disease, so many lives affected, not just the PD patient, but spouses, mothers, fathers, grandparents, siblings, care givers, doctors, researchers, teachers, exercise specialists. What a toll! And it goes on, and on, and on. No sudden death. An endless progression. Spiraling downward, downward, downward.

With so much at stake, so many lives, how can we overlook the One Great Power of the Universe, Almighty God? Take a look at the suffering of the Pope. Inspiration, yes, but consider, could I undergo such public exposure while I am sinking with an incurable disease?

Maybe yes - maybe no. Throughout this book you may have noticed that I want to be a "real person" again. Well, that's my

. . . DREAM NUMBER ONE

— 10 — HOPE FOR THE FUTURE

So I pray daily for help from God, to overcome my disabilities, to ease the burden on those loving people who help me, to assist my wife and family, for a cure, if not for me, for PD victims in the future.

You see, I'm not so strong, or smart, or hard-working - I need God's help. And HE IS GIVING ME HELP. This gives me hope for moving on with Parkinson's disease. No one can predict the future, but we can try to shape the future.

First, do the best you can with what you have. You know better than anyone what your strengths and weaknesses are. Build on strengths, improve weaknesses.

Get help from people who can help. You will be better equipped for the future, whatever it may bring, if you have support and help. Above all, get a really good doctor, like Dr. Rao. Find an experienced Personal Trainer (See Appendix IV, for Maribel Bleeker's resumé).

Do not become discouraged. There will be new breakthroughs in the treatment of PD. The literature is filled with all kinds of studies, trials, research efforts, clinical evaluations and even some possible cures under investigation.

No matter how severe your symptoms are, go one day at a time, and work on improving your future. One way to do this is to work at a dream. Make it happen. You can do it. Look for a dream that can make you happy.

Subscribe to a neurology publication to stay in touch with developments. An example is Neurology Now, free to neurology patients. Visit www.neurologynow.com.

If you are getting help from a care provider, help them in return, by being a good patient, smile and laugh a lot.

Whether or not you are spiritually motivated, pray for help.

Hope for Tomorrow

Parkinson's disease can be a long journey; it can be painful, frustrating, and frightening; it can change lives in a family when someone has PD.

But - my friend, there is a TOMORROW! Do you understand? This disease is not fatal. We PD patients have a chance. There IS a TOMORROW! Maybe tomorrow will bring a new study by some really sharp researchers who will find a way to cure Parkinson's disease. It could happen, and I am convinced that eventually it WILL happen.

So what will the future bring? It will give us hope and progress. And above all, with God's help it will bring a cure.

Don't worry. Get busy.
Work hard. Dream a lot.
Make your dreams come true.
And remember,
the Parkinson's beast will be slain

FOR SURE!
DON'T GIVE UP HOPE

The difference between a successful person and others is not a lack of strength, not a lack of knowledge, but rather a lack in will.

Vince Lombardi

EPILOGUE

I wish that I could announce that medical science has found a cure for Parkinson's disease, but that is not the case. Not yet, at least. Someday—when I do not know—there will be a news conference in which triumphant scientists and doctors will tell the world that they found a cure for PD. Tears of joy will flow freely. Families will hug their PD patients. Governments will follow suit with free treatments so that everyone with PD can be cured. The long nightmare will be over — for good.

In the meantime, we must make do, in any way we can afford, to live with PD. I anticipate that my readers will make several valid points regarding my advice in the book. First, my treatments, involving travel several times a year, expensive medications, and costs of a personal trainer in my home, are beyond the financial means of some people with PD.

Second, it also is true that my case of Parkinson's is less severe than more extreme cases, and therefore, these patients cannot take advantage of some of the advice given in the book.

Third, it is easy for me to speak of goals and ambitions, since I seem to be a lucky patient who responds so well to medical and physical treatments.

Even granting these points, should these issues be allowed to rule the day? I think not. Remember your internal computer, your BRAIN. Let's use it to devise solutions to these problems.

1. Use the Internet to learn about medication prices from various sources. There are offers on TV and the Internet for free or low cost drugs. If you do not have a computer with Internet access, ask a relative or friend for help, or a local library. Be a smart shopper for medication. Do you have a drug plan that can help with some costs? Does your state have assistance programs for low income persons needing prescription drugs? Some drug manufacturers offer low income assistance.
2. Get help from family or friends to undertake an exercise program. Even a few hours a week of exercise is beneficial. Check on eligibility of Senior Citizens for Medicare coverage of physical therapy as an outpatient. Some gyms cooperate with the free Silver Sneaker program for Seniors with Humana Medicare supplement insurance.
3. Do some research on specialists who are trained and experienced in Parkinson's disease. Please refer to Appendix V (page 144) for information on finding a specialist in your area. Look for a neurologist who works primarily with Movement Disorder and Parkinson's disease patients.

Have courage. Have a dream. Work hard. Be resolved to beat PD. No matter how bad your symptoms, fight back. Tomorrow is another day. A cure is coming. I am convinced. Sooner or later, *a cure is coming!*

In the meantime the world still revolves once a day. This is to say that life must go on, regardless of the misfortune of PD entering one's life. Parkinson's can strike anyone in any position in life. I can empathize with the newly diagnosed person who says "Why me?" This diagnosis is mind shattering. And it is incredible that some in the medical profession, including specialists who should know better, can be insensitive in their handling of PD patients. Shame on them.

So you pick up a book like the one you are reading now. It suggests the need for a constant attack on PD, including medication, PD specialists, physical therapy, counseling, and participation in support groups.

But what about the expense, the time needed to fight the disease, the slow progress in getting symptoms under control? One might ask an "experienced" PD patient, was it worth it? The years of hard work, the money, the time—*WAS IT WORTH IT?*

If you ask me, it was absolutely worth it. Several people who know me recently told me, very sincerely, that they would never suspect that I had a disease. It took a lot of time and money to hear those words.

But now, I can truthfully say that I am REAL PERSON again. My next goal is to continue to suppress symptoms and go backwards from Stage 3 to Stage 2.

Remember, this is a war, and I am DETERMINED to be triumphant!

God bless us all, and help us, the Victims of Parkinson's Disease, as we all struggle against this foe.

APPENDIX I

PARKINSON'S DISEASE – AN EXERCISE PROGRAM TO RESTORE AND MAINTAIN WELLNESS

My story would not be complete without some mention of an exercise program that has been successful in restoring a great deal of my physical condition. This program has been effective when used in conjunction with medical treatment.

Let me begin by describing the bottom that I reached just prior to a diagnosis of Parkinson's disease. PD is insidious because it can do so much damage to the body before it is diagnosed and treated. At least that's what happened to me. The descent toward physical collapse happened in my case so slowly that I didn't realize what was happening.

Over several years, I gradually lost strength in my arms and legs, back, hands and feet, that is, just about everywhere. This loss was at first very subtle. My joints lost range of motion and became stiff. It was just arthritis, I thought. My posture worsened, as my upper torso began to pitch forward. I lost over three inches of height. I lacked coordination, by which I mean that my walking and other movements were jerky, and I was clumsy. I had trouble getting out of a chair. I tended to stumble when walking on uneven surfaces.

My hands became weak and unreliable, and my manual dexterity was gone. My writing was cramped and hard to read. Simple tasks that required the use of my hands became virtually impossible. It was really difficult to get dressed. I could not seem to get my shirt buttoned without help. I absolutely could not tie a necktie. Eating presented a terrible challenge – I could not cut my meat, and I dropped food all over myself.

My sense of balance deteriorated so that I felt unsteady just standing still. My voice became weaker and my wife had to ask me to repeat a lot. My endurance in repetitive tasks was poor. I became exhausted just on a short walk in my neighborhood.

Jeanne urged me to seek medical help, which I did. Ultimately I was fortunate to have Dr. Jay Rao, M.D., a Specialist in Movement Disorder Diseases at Ochsner Clinic in New Orleans, Louisiana, to prescribe the medical treatment of my case.

I am not a medical doctor nor do I have a license to dispense medicine or offer advice on treating PD or any other disorder. But I have another unique qualification to speak about PD - I am a PD patient in Stage 3 who can report what was done to recover much of my life energy with an exercise program designed and directed by Ms. Maribel Bleeker, a Health Fitness Specialist.

MANUAL DEXTERITY AND FINE MUSCLE CONTROL

Unless you experience PD, you may not recognize that there are so many important muscles in your body. Just about every part of your body depends on some size and type of muscle to function properly. We are trying to keep my hands as strong as possible with hand and finger exercises. Other fine muscles include the eyelids, which may not blink enough, so the eyes become dirty. The muscles that control the position of the eyes and their focus also can be exercised.

Hand and Finger Exercises

VOICE, SWALLOWING AND CHEWING

Did you know that there are eight muscles inside your tongue? These muscles must work properly not for just licking postage stamps, but to TALK and CHEW and SWALLOW your food. And what happens if you lose the functioning of these muscles? Then you can't talk and may choke on your food. Sad, but there apparently are PD victims who have waited too long to keep these muscles working. But my speech and eating are JUST FINE! How did I do that? *EXERCISE.* Maribel recommended that I see a Speech Therapist at the local hospital out-patient section. I was given a set of tongue exercises. We also worked on the VOCAL CORDS. Guess what is behind the successful operation of the vocal cords – yes, muscles. I was given a series of exercises for my vocal cords and to keep my enunciation and fine speech working. We also practice enunciating certain words, like "CHEEZEBURGER," which exercises both the front and back muscles in the tongue. We incorporated these exercises in my exercise program.

FACIAL EXPRESSIONS

I used to wear the "Parkinson's Mask." If you have PD or know someone who does, you know what I mean by MASK. The face is blank, no expression, eyes unblinking, entire face sagging, lifeless—DEAD. This poor soul might as well be dead. I recall being in a supermarket checkout line wearing one of these dead faces. The teenage girl bagging the groceries stared at my face, like I was a Zombie to be feared. With proper PD medication and exercise of the facial muscles, the face can be restored to a fairly normal condition. Maribel has me give her a repetitive "big smile," followed by a "surprise" face. She does the same back to me and we both end up laughing (which makes for a better smile from me).

UPPER BODY STRENGTH

The exercise program devotes good effort toward maintaining and improving the muscular strength of the arms, chest, back, shoulders, and abs. We employ several tools that are appropriate for a home setting, including lifting weights, push-ups, and resistance bands. Two examples are shown below:

Shoulder Press with Resistance Band

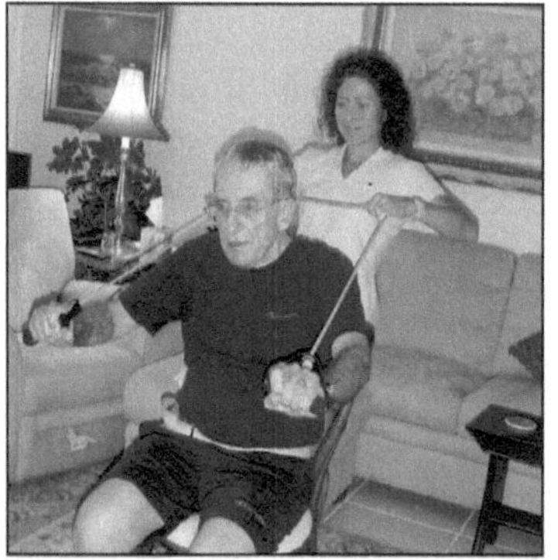

Biceps Curl with 5 lb. Dumbbells

RANGE OF MOTION

One of the annoying aspects of PD is the loss of motion of the joints. For example, you drop a pen on the floor and, if you don't have PD, you just bend over and pick it up. When the PD patient tries to bend over, the pen is always just out of reach. Here I am performing motion exercises to keep the joints open by swinging a two-lb. weight through a wide arc.

Range of Motion Exercise-Left Shoulder

BALANCE

The PD patient should constantly think, "I must not lose my balance and fall." The exercise program concentrates on balance practice. One routine involves walking while turning my head side to side. Another exercise uses a thick balance pad to step up and down with alternating left and right feet. Still another has me standing with my eyes closed and both feet together. My balance has improved so much that I now balance on a more challenging "BOSU" balance trainer.

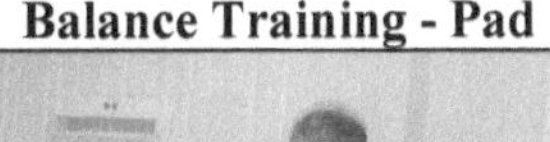

Balance Training - Pad

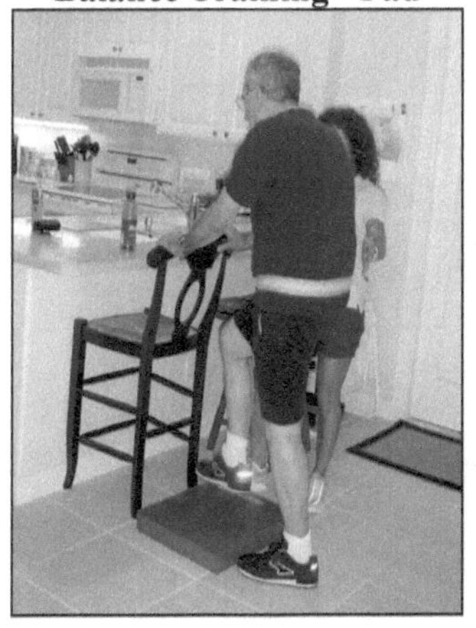

Balance Training - BOSU

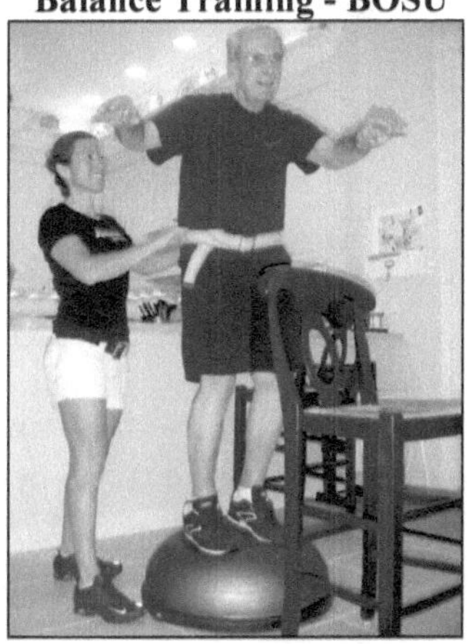

LEG EXERCISES

No wheelchairs allowed! Use them or lose them – your legs, that is. There are very important muscles in the legs that need to be exercised frequently. First, simple walking, heel-toe, stand up straight, head up, arms swinging at your side, again and again. For the PD patient, this exercise takes real concentration. Now, march, knees high, control-control, again and again. Now change again – side step, first left, then right, again and again. Do you have trouble getting out of a chair? Let's do 15 "sit to stands." Sit in a straight-back chair, then stand up, sit down, and see if you can do this 15 times in less than 45 seconds. (My best time is now 17 seconds!) I do this exercise three times a session. My reward is that I can walk into a restaurant, sit down at a table, eat my dinner, pay the bill, stand up and walk out, *JUST LIKE A REAL PERSON!*

STRETCHING

I have left an important subject to last. What does a person about to workout in a gym do first? That's right – warm up and stretch muscles to properly prepare for their exercise program. Don't skip this step, or you might injure yourself. I am seen stretching my legs in the photo below. I always roll my feet and ankles, do "heels up - toes up" while sitting, and stretch arms and shoulders.

Pre-Warm Up and Stretch

Between the medical treatment of my PD symptoms and the exercise program described above, I have experienced a dramatic improvement in my physical and mental wellness. Give it a try. Be positive. Don't feel sorry for yourself, get to work – EXERCISE!

APPENDIX II

RESOURCES FOR PARKINSON'S DISEASE

National Parkinson Foundation
1-800-327-4545
www.parkinson.org

Parkinson's Disease Foundation
1-800-457-6676
www.pdf.org

Michael J. Fox Foundation for Parkinson's Research
1-800-708-7644
www.michaeljfox.org

American Parkinson Disease Association
1-800-223-2732
www.apdaparkinson.org

NOTES

APPENDIX III

PARKINSON'S DISEASE MEDICATIONS

The medication regimen prescribed for the author included the following prescription drugs and doses:

REQUIP (ropinirole)

- Type - dopamine agonist
- Dose - three 3 mg tablets per day (daily dose 9 mg)
- Drug manufacturer - **GlaxoSmithKline**
- Contact information - www.gsk.com
- Efficacy - moderately high
- Side effects noted - nausea, light headedness, drowsiness, depression

STALEVO 100 (25 mg carbidopa, 100 mg levodopa and 200 mg entacapone)

- Type - Dopamine Replacement
- Dose - six tablets per day (every 4 hours, 24/7)
- Drug manufacturer - **Novartis**
- Contact information - www.stalevo.com
- Efficacy - high
- Side effects noted - not significant at prescribed dose

NEUPRO * (rotigotine transdermal system - 24 hr. patch)

- Type - dopamine agonist
- Dose - delivers 4 mg/24 hours
- Drug manufacturer - **Schwarz Pharma**
- Contact information - www.neupro.com
- Efficacy - high
- Side effects noted - Irritation of skin after repeated application of the patch in one area

---

** As of the publication date, Neupro had been recalled due to production problems related to crystals impeding the delivery of the medication.*

NOTES

APPENDIX IV

MARIBEL BLEEKER

Health Fitness Specialist

Maribel Bleeker is the owner and CEO of Maribel Bleeker Fitness, LLC, and has 10 years of experience in the fitness industry. She holds a Bachelor's Degree in Exercise and Sports Sciences from Florida International University and works as a Health Fitness Specialist providing personal training for individuals in the Palm Beach County area. Maribel enjoys giving lectures to community groups on topics such as fitness for everyone and fall prevention. Her passion is to coach and teach people how to make healthier choices in their lives through fitness and healthy food options.

Ms. Bleeker loves fitness and always looks for a challenge. She has been an avid runner for over 20 years and enjoys competing in races from 5ks to marathons. Maribel works with individuals ranging in age from 8 to 80. She specializes in Adventure Boot Camps, Women's fitness and fat loss, Senior fitness, ***over five years experience in Parkinson's disease and fitness,*** Positive lifestyle changes and Online training to name a few. Maribel strives to empower people to believe in their potential and in their ability to reach their goals. "Once we can understand our goals and define them it helps to give direction toward reaching them." She is a firm believer that fitness is for everyone and that it is never too late to get started.

Education:

- B.A. in Exercise and Sports Sciences – Florida International University

Certifications:

- Adventure Boot Camp Instructor – National Exercise & Sports Trainers Association (NESTA)
- Health Fitness Instructor – American College of Sports Medicine (ACSM)
- Strength Conditioning Specialist – National Strength Professionals Association (NSPA)
- Fitness Nutrition Coach – NESTA
- Spin – Cycle Reebok
- Body Pump Instructor – Body Training Systems
- Exercise and Parkinson's Disease – DSW Fitness
- Designing Fall Prevention Workshops – DSW Fitness

1-561-315-1239
www.NpbBootCamp.com
e-mail: maribel@NpbBootCamp.com

NOTES

APPENDIX V

FIND A DOCTOR– PARKINSON'S DISEASE

Groups that maintain lists of Movement Disorder and PD specialists

- National Parkinson Foundation, 800-327-4545, www.parkinson.org

  NPF maintains an online resource called *Decide-PD.* This "Find a Doctor" directory provides detailed information on PD specialists by state. Categories include education, board certification, type of practice (academic vs. private), percentage of patients seen for PD/movement disorders.

- American Parkinson Disease Assoc., 800-223-2732, www.apdaparkinson.org

  APDA maintains an Information Referral Center that offers persons with PD and caregivers education, support, public awareness and referral services. Referral Centers are listed by state and include the Center location, phone number and the e-mail address of the Referral Center Coordinator.

- Parkinson's Disease Foundation, 800-457-6676, www.pdf.org

  Call for assistance or send an e-mail to info@pdf.org. Online support includes articles on choosing a doctor and questions to ask in the selection process.

www.ingramcontent.com/pod-product-compliance
Lightning Source LLC
Chambersburg PA
CBHW030753180526
45163CB00003B/1001

9780557064991